W9-BNQ-450

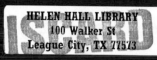
May 19

GRAVITY'S CENTURY

GRAVITY'S CENTURY

From Einstein's Eclipse
to Images of Black Holes

Ron Cowen

Harvard University Press
Cambridge, Massachusetts
London, England
2019

First printing

Library of Congress Cataloging-in-Publication Data
Names: Cowen, Ron, 1957– author.
Title: Gravity's century : from Einstein's eclipse to
images of black holes / Ron Cowen.
Description: Cambridge, Massachusetts : Harvard University Press, 2019. |
Includes bibliographical references and index.
Identifiers: LCCN 2018052381 | ISBN 9780674974968 (alk. paper)
Subjects: LCSH: General relativity (Physics) | Gravity. |
Quantum theory. | Astrophysics. | Gravitation.
Classification: LCC QC173.6 .C36 2019 | DDC 530.11—dc23
LC record available at https://lccn.loc.gov/2018052381

To Cathy and Julie for all the joy
and love they bring to my life

Contents

GRAVITY'S CENTURY

INTRODUCTION

For the monster at the Milky Way's heart, it was a wrap.

On April 11, 2017, a team of astronomers completed five nights of observations with a network of radio telescopes that collectively acted as an antenna with a diameter the size of Earth. Stretching from Hawaii to the South Pole, eight groups of telescopes had spent those nights staring at something nobody had ever seen before, although visionary scientists and sci-fi aficionados alike had been trying to imagine it for decades: a black hole.

Or, more precisely, the edge of a black hole—an outer shell of gravity so strong that nothing can escape its grasp, not even light. Once something crosses that boundary—what scientists call an event horizon—it's gone from our universe forever. But even though it's disappeared, the stuff inside a black hole participates in a distortion of space and time that defies logic and math and that, for anyone curious about nature at its most extreme, must be seen to be believed.

Belief alone, however, did not motivate the creators of the Event Horizon Telescope. Understanding did, because to begin to understand a black hole is to begin to understand the universe anew—to enter an era of exploration that Albert Einstein himself denied would ever happen, even though he, more than anyone, had made it possible.

A century earlier, in 1915, Albert Einstein finalized his general theory of relativity. He alone had realized the full import of a law of physics that Galileo had discovered nearly 300 years earlier: all objects, regardless of their mass or composition, fall at the same rate in a gravitational field. Making that law the heart and soul of a revolutionary theory about the acceleration of objects, Einstein had forged a new way of thinking not just about gravity but about the universe.

He erased the idea of gravity as a force and refuted long-held notions of space and time as featureless, silent spectators to the comings and goings in the universe. Space-time was as malleable as putty, shaped by the presence of mass and energy. A body did not fall because Earth tugged on it; rather, Earth's mass and energy curved the surrounding space-time in such a way that a passing body would inevitably have its path bent toward the planet. The same mutual influence applied to any two objects in the universe. Even light was subject to this law of nature: if it passed near a massive-enough body—the Sun, for instance—its path, too, would bend.

During a solar eclipse on May 29, 1919, with the ravages of World War I still fresh, two teams of British astronomers trekked to Brazil and the west coast of Africa to test the strange new theory of gravity by the German-born Einstein. As the Moon inserted itself between the Sun and Earth for six minutes and 51 seconds on May 29, 1919—one of the longest solar eclipses of the twentieth century—the teams photographed the stars that came into view as brilliant day turned to sudden night. When the observers went home and compared images of the same stars when they were not near the Sun, they found just what Einstein had said: the Sun's heft bends starlight. For this prediction alone, Einstein became a celebrity overnight, his theory grabbing headlines around the world.

Gravity's century had begun.

These two experiments—the eclipse expeditions of 1919 and the Event Horizon Telescope observations of a century later—bookend an era unlike any other in the history of science.

One hundred years ago, when Einstein formulated his general theory of relativity, the universe seemingly consisted of a single galaxy; today we know not only that the universe has at least 100 billion galaxies, but that it is expanding, ballooning at a faster rate every second. And over the past century astronomy has grown from the study of the narrow optical band of the electromagnetic spectrum, visible through individual telescopes, to the whole range of the electromagnetic spectrum, from microwaves through gamma rays. By the start of the twenty-first century, astronomy had extended even beyond the electromagnetic spectrum: mysterious, invisible entities known as dark matter and dark energy are now known to make up 96 percent of the cosmos.

Anyone trying to make sense of these discoveries owes a debt to general relativity. But no phenomenon is quite as counterintuitive as a black hole. The idea was a natural outcome of general relativity, as some theorists realized as soon as Einstein had devised his theory.

If an object was massive and dense enough, wouldn't space-time become so distorted that it would close in on itself? Not only would light passing close to such an object be bent; if it passed too close, it would fall into the gravitational trap and never escape.

Einstein never liked the idea of black holes; it made his elegant equations blow up and lose their meaning. For decades, he and other physicists could afford to ignore the concept.

But then another kind of revolution arrived—this one in telescope technology. Observations beginning in the early 1960s revealed that compact beacons of radiation from the distant reaches of the universe were outshining entire galaxies, and that stars were whipping around galactic centers at staggeringly high speeds. The enormous energies and furious velocities betrayed the presence of unseen gravitational hulks at the cores of galaxies. Black holes, light-guzzling gravitational maws in space-time, had become real.

Soon theorists began to take a strong interest. They realized that black holes were a crucible for marrying the world of the very tiny—the realm of quantum theory—with the realm of extreme gravity, where general relativity rules supreme. That was a marriage Einstein had spent decades trying to forge but never accomplished.

And when astronomers realized that new radio telescope technology could allow them to image the actual event horizon of a black hole—well, who could resist?

The Event Horizon Telescope collaboration chose to target two black holes in particular. One of those gravitational monsters was Sagittarius A*, which churns at the center of our Milky Way galaxy and has a mass 4 million times that of the Sun. The second, possessing a mass some 1,000 times greater, occupies the core of M87, a galaxy 54 million light-years distant. Together they may provide science with a crucial test of Einstein's general theory of relativity: determining how well its predictions match observations in the most extreme gravitational environment known in the universe.

At 11:22 A.M. Eastern Time on an otherwise nondescript spring day in 2017, the Event Horizon Telescope recorded its last photon of the observing season. The researchers knew they still had many months of work ahead of them. They would be analyzing a data set equivalent to the storage capacity of 10,000 laptops. At the same time they would also be preparing for a second observing run the following year. But now, for this one moment, rather than focus on how far they had to go, they could pause to appreciate how far they'd come.

One astronomer blasted the triumphant chords of Queen's "Bohemian Rhapsody." Another cracked the seal on a bottle of fifty-year-old Scotch. But while the immediate cause of the revelry was the completion of an experiment that covered the width and breadth of Earth, the historical context was even broader. As the revelers well knew, theirs was a celebration a century in the making

1

GENESIS

TWO MONTHS. That's all the time Albert Einstein had in September 1907 to write the first definitive review article on his special theory of relativity, which he had published two years earlier. Einstein, who at twenty-eight was still hunting for a university teaching position while working as an examiner at the Swiss patent office in Bern, embraced the opportunity to summarize his controversial work for the prestigious *Yearbook on Electronics and Radioactivity*. But he twice queried the journal's editor about when the article was due.

Einstein's concern was understandable. To support his wife and three-year-old son, he put in eight-hour days, Monday through Saturday, at his desk on the third floor of the new Postal and Telegraph Building, where he judged the merits of proposed electrical inventions and other contraptions. He worked so efficiently—patent office director Friedrich Haller held him in high esteem—that he found time during the workday to pursue his research in theoretical physics.

When Einstein completed his manuscript in November 1907, it did much more than explicate his original work. His review contained the seeds of a brilliant, broader, and much stranger theory of relativity that would forever change how humans perceive the cosmos.

Einstein began the article by recapping his 1905 paper, in which he had radically reimagined the classical notion of relativity described by Galileo Galilei. In Galileo's 1632 book *Dialogue Concerning the Two Chief World Systems,* the Italian scientist and inventor asserted that the behavior of objects is identical whether they are at rest or moving at a constant speed.

Galileo features three characters in his book—Salviatus, a stand-in for Galileo; Sagredus, an intelligent layperson; and Simplicius, who is none too bright. Together, they investigate whether the laws of motion governing objects will appear any different if an observer is at rest or moving at constant speed.

First, Galileo asks us to consider a ship anchored at a dock. If someone drops a stone from the ship's mast, it would strike the part of the deck that lies at the base of the mast. That seems perfectly obvious to everyone, whether they are on the ship or standing on the dock.

Now consider the same ship, Galileo says, but this time it's moving at a constant speed on the water, let's say 10 meters per second. Repeat the same experiment—if someone drops a stone from the mast on the moving boat, where will the stone land? If the stone takes one second to drop, wouldn't the stone land 10 meters behind the mast, since the ship has moved 10 meters forward during that second? That's the answer Simplicius would give, and it may seem correct. But it's wrong.

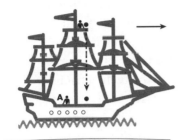 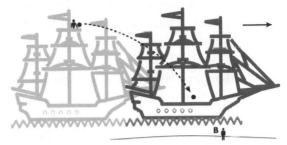

Two perspectives of Galileo's description of someone dropping a cannonball from the mast of a ship moving at constant velocity. To the person atop the mast, who moves with the ship, the cannonball appears to drop straight down (*left*). To someone at rest on shore, the cannonball appears to follow a diagonal path to the bottom of the mast (*right*), since the ship has moved at constant speed during the time it took for the cannonball to fall. But both observers agree that the cannonball ends up at the bottom of the mast. *(Courtesy Kristen Dill.)*

The stone would still land at the base of the mast, just as when the ship was at rest. The laws of motion are the same whether the boat is at rest or moving at a constant speed.

To the sailor who dropped the stone from the top of the mast, the stone falls straight down. If you were standing on the dock, you'd agree about where the stone hit, but you'd perceive the stone as following a diagonal path, because from your point of view the stone has some forward motion, the same forward motion as the boat.

Regardless of the different path the falling stone appears to take for each observer, the laws of physics and motion are the same for those two observers. In fact, said Galileo, if you were a passenger in a windowless cabin below deck and the boat was moving at a constant speed, no experiment could tell you whether the boat was in motion or standing still. Fish swimming in a bowl or butterflies flitting through the cabin would always move in the same way.

It was Einstein's masterful idea to rethink and extend Galileo's classical principle of relativity by, in effect, replacing the falling stone with a beam of light.

Light was something Einstein had been thinking about since he was a child. When he was twelve or thirteen, a friend gave Einstein a book by science fiction writer Aaron Bernstein, who took his readers on a journey into deep space. You don't go by boat or speeding train; Bernstein asks you to imagine riding alongside an electric current as it races through a telegraph wire.

Einstein was thrilled by the imagery. And when he was sixteen, enrolled in an avant-garde school in Aarau, Switzerland, that encouraged visual thinking, he imagined an even more fantastic voyage—riding alongside a light beam. What would a light beam look like if he could travel fast enough to catch up to it?

The light wave, he initially figured, would look stationary, immobile, just as a speedy runner would look if you could keep pace with that person. But the idea of a motionless light wave not only violated everyday experience but also contradicted what Scottish physicist James Clerk Maxwell had revealed about light and its connection to electricity and magnetism. Maxwell's set of equations demonstrated that electricity (the force between charged particles) and magnetism (for example, the attraction of two bar magnets) are not separate phenomena but two facets of a single entity called electromagnetism. From his equations, Maxwell also found that when oscillating electric and magnetic fields are set at right angles to each other, they generate a wave that travels at exactly 299,792 kilometers per second. That's the speed of light. A light beam is an electromagnetic wave.

But what was the speed relative to? Physicists at the time decided the speed must be relative to some kind of medium, the same way sound waves need a medium, like water or air, in which to travel. Scientists called the medium for light an ether but could find no evidence for it. Einstein knew better—there was no need for an ether because the speed of light was not relative to any reference frame or medium. Einstein demanded that the laws of physics, including Maxwell's equations, must be the same, and yield the same answers, for all observers moving at constant speed relative to each other. Since Maxwell's equations predict a particular value for the speed of light, that speed should be the same for all observers in uniform motion. It was *always* 299,792 kilometers per second, he asserted. Which meant you could never catch up with a light wave.

On the face of it, that's a crazy idea. The slow speeds we're familiar with are additive—if I'm in a train going 15 miles per hour and look across the track and see a train going 10 miles an hour in the same direction, my speed relative to the other train is 15 – 10 = 5 miles per hour. So if you could travel really fast, at nearly the speed of light, wouldn't you see a passing light wave travel more slowly than someone at rest would? Einstein—and numerous experiments—said no, you'd see the light wave race past you just as fast as if you were standing still. The speed of light does not vary. And on top of that, nothing can travel faster than light.

Speed is measured as distance divided by time (miles per hour or meters per second). For the speed of light to remain constant, distance and time have to change.

Let's go back now to Galileo's shipboard experiment, using a beam of light instead of a stone. On a boat that's moving at a uniform speed across the water, shine a flashlight down the mast, and it will strike the deck at the base of the mast. The observer on the dock agrees with that. But from her vantage point on the dock, if she had a precision measuring tool, she would see the light travel a tiny extra distance, the distance the ship has moved in the time it took the light to reach the bottom of the mast.

But the speed of light, which is measured in meters per second—again, distance divided by time—is a constant. So if the observer on the dock finds that light traveled an extra distance, the only way its speed can remain constant is if the light also took a longer time to travel.

Time, therefore, is *not* immutable. The duration of time—measured as the ticks of a clock—is different for observers who move at different speeds. Each sees the other's clock slow down. Even more strangely, distance is not absolute either; it appears to contract in the direction of motion.

Why didn't people notice this sooner? Because the changes in space and time are minuscule unless you're moving at close to the speed of light. Or as my physics professor at New York University, Larry Spruch, used to say, if we all had relativistic (fast-moving) toys growing up, we'd understand intuitively that space and time are not absolute.

But it turned out that a new quantity—space-time—did stay the same. The mathematician Hermann Minkowski, who had thought of Einstein as "a lazy dog" when he taught him at the Federal Institute of Technology in Zurich a decade earlier, was the one who realized that space and time were on equal footing. He re-formulated special relativity as a four-dimensional theory, with time added as the fourth dimension to the usual three dimensions of space. "Henceforth space by itself, and time by itself, are doomed to fade away into mere shadows, and only a kind of union of the two will preserve an independent reality," Minkowski declared in 1908. (For more on space-time, see "Deeper Dive: Space and Time, a Perfect Union.")

That was the essence of Einstein's special theory of relativity. But as he continued to write his 1907 review article, Einstein was not satisfied. His theory applied only to observers who moved at constant velocity with respect to each other. Someone moving at constant velocity not only travels at a constant speed, but also does not change direction. What about observers who were slowing down or speeding up or changing direction? Speeding up or slowing down meant acceleration, and acceleration could be accomplished by applying a force—like the gravitational force described by Isaac Newton. But Einstein could not fit Newton's laws of gravity into his new space-time picture.

Newton's theory of gravitation beautifully describes the motion of the planets and the shapes of their orbits—elongated circles, known as ellipses—around the Sun. The theory even predicted the existence of Neptune before it was discovered. But Newton's law of gravitation has a flaw even Newton acknowledged. Gravity, in his theory, is a force that acts instantaneously, no matter how far apart two masses lie. Although it takes eight minutes and twenty seconds for light emitted by the Sun to reach Earth, Newton's gravity communicates the Sun's gravitational pull without any delay, defying the cosmic speed limit that nothing can travel faster than light.

But then, around the time he was completing the review, "the happiest thought" came to Einstein. It happened while he was sitting in his chair at the patent office in Bern. If a man fell off the roof of a building, Einstein realized, he would not feel his own weight. *He would not feel the force of gravity.* The thought startled him.

Although it would hurt like hell when the man hit the ground, during the time he was in free fall, he would not experience gravity. And if, while he was falling, the man dropped a ball or took the house keys from his pocket and released them, those objects would hover in midair, floating alongside him. The man would feel as if he were at rest.

Einstein's *Gedankenexperiment,* or thought experiment, led him to another realization. An observer can replace the downward pull of a uniform gravitational field, at least in his immediate vicinity, with a constant acceleration upward—that is, increasing speed at a constant rate. The two are equivalent. "It is impossible to discover by experiment whether a given system of coordinates is accelerated, or whether . . . the observed effects are due to a gravitational field," Einstein said.

The idea bears repeating, so let's put it another way. Imagine you're in an elevator in outer space, away from all gravitational forces. Some agent is tugging at the top of the elevator so that it accelerates upward at a rate of 9.8 m / sec². Your feet press against the floor of the rising elevator in the same way that your feet press against Earth's surface due to gravity. Should you drop a ball, it would accelerate to the floor, just as it would on Earth. You could not tell whether you were accelerating upward or if you were at rest on Earth. (The idea is reminiscent of Galileo's passenger shut in the windowless cabin below deck who cannot tell whether the ship is at rest or moving at a uniform speed.)

Instead of just sticking with the idea that gravitation and acceleration are equivalent, Einstein made an even stronger declaration: that all the laws of nature are identical whether you are in a static, uniform gravitational field or in a reference frame, like an elevator, that is uniformly accelerating. The results of any experiment conducted in a uniform gravitational field and in the elevator are the same—whether you're doing the Macarena, juggling balls, performing a somersault, or shaking a cocktail. This is the principle of equivalence, and it has far-reaching implications for the nature of space, time, and the universe.

But why should this hold true? For that we can look back to Newton's laws of motion and gravitation. Newton stated that force (F) is equal to mass (m) times acceleration (a): $F = ma$. Here m is the inertial mass, referring to the property of an object that resists change to the object's motion. It's the inertial mass you have to overcome to push a car that's at rest, or to stop it once it's moving.

Separately, Newton calculated the gravitational force between two bodies, which is proportional to the product of each of their masses divided by the distance between them squared ($F = m_1 m_2 G / r^2$). Here m_1 and m_2 are the gravitational masses (a measure of how strongly the objects are attracted to a gravitational field), G is the gravitational constant, and r is the distance between the objects. The two types of masses—inertial mass and gravitational mass—turn out to be exactly equivalent.

This is why a lead brick and a ball of cotton fall at the same rate in a vacuum (in the absence of air resistance). In Galileo's apocryphal experiment at the Leaning Tower of Pisa, he dropped two balls of different weights. Assuming no effect of air resistance, they would have hit the ground at the same time. (For more on the history of such experiments, see "Deeper Dive: Testing the Equivalence Principle before Einstein.") Gravity is an equal-opportunity interaction—it affects all objects in the same way, regardless of their mass, size, shape, electric charge, or other properties.

It did not have to be that way. In theory, an object with more gravitational mass might fall more rapidly to Earth than a body with less. In that case, all objects would not fall at the same rate in a gravitational field and the effects of gravity could not be replaced by uniform acceleration. For example, if you were in free fall and dropped a small ball that weighed much less than you, the ball would not hover next to you—you would fall faster.

But why should gravity, something you feel while motionless on Earth's surface, have this intimate link to accelerated motion? Einstein intuited that the deep connection between gravity and acceleration was related to some intrinsic property of space-time itself.

For the next few years after his 1907 review article, Einstein's attention was captivated by other puzzles in physics. But in 1911, he returned to gravity with yet another variation of his thought experiments.

A beam of light enters an elevator that is ascending at constant acceleration. To a stationary observer outside the elevator, the light follows a straight line across to the opposite wall of the elevator cabin (*left*). But for someone inside the accelerating elevator, the path of the light is not straight. By the time the light has traversed the width of the elevator cabin, the cabin has moved upward at a faster and faster rate (indicated by the three elevator panels). To the person inside the elevator, the light appears as if it has bent downward, because it enters the elevator at a point above her head but strikes the cabin's opposite wall near the elevator's floor (*right*). Because of the equivalence between constant acceleration and a uniform gravitational field, this means that gravity bends light. (*Courtesy Curt Suplee.*)

We're once again back in an elevator in outer space, away from any gravitational forces. Hoisted by a crane, the elevator moves upward at constant acceleration. A tiny hole has been drilled in one of the walls of the elevator cabin. Suppose someone outside the elevator shines a light aimed horizontally, through the hole, the moment the elevator begins ascending.

How does the light beam travel? From the point of view of the crane operator, who is outside the elevator and not accelerating, the beam traverses the width of the elevator in a straight line.

The passenger in the elevator has a different perspective. During the time it takes the light to cross from one side of the elevator to the other, the elevator cabin will have moved upward.

To the passenger, who does not know she is accelerating upward, it looks like the beam of light has moved closer to the floor during its travel—in other words, that it has fallen, in the same way that anything else in the elevator that is dropped from rest would fall. To her, it appears that the light beam has bent downward.

But an elevator that is uniformly accelerating is equivalent to a uniform gravitational field. So by the equivalence principle, we must conclude that gravity bends light!

Could that really be true? What kind of evidence could demonstrate it? The bending of light due to Earth's gravity would be so slight that it would not be noticed. But in 1911, Einstein realized that the Sun's gravitational field should be strong enough to noticeably bend passing starlight. The effect could in theory be observed during a solar eclipse, when the Sun's brilliant disk is dimmed. His prediction was validated by observations during the solar eclipse of 1919. It became the most celebrated confirmation of Einstein's theory (see Chapter 3).

Bending light is not gravity's only sleight of hand. Recall that it also makes clocks run slower. Another thought experiment illustrates this. Back one last time to the accelerating elevator. Someone on the floor of the elevator sends a flash of light once per second to an observer perched atop the moving cabin. The light flashes are like the ticks of a clock—each time another flash goes off, another second has elapsed, according to the person on the elevator floor.

As each flash travels to the observer at the top, the elevator cabin is moving upward at a faster and faster rate. It therefore takes a longer amount of time for each light flash to reach the top. So the observer atop the elevator measures an interval between flashes that is longer than one second, the rate at which the person on the floor of the elevator is sending them. The observer on top concludes that time passes more slowly for the person on the floor of the elevator.

Applying the equivalence principle, what's true for acceleration must hold true for gravity. That means that intervals between
flashes, or the ticks of a clock, are faster for someone on a mountaintop, farther away from the center of Earth and its gravitational
pull, than for someone at sea level, where the gravitational field
is stronger. Gravity slows time.

There was one more thought experiment that crystallized the
radical role space and time would play in Einstein's new theory
of gravity and in the mathematical tools he would need to master
in order to arrive at a final theory. The experiment was first suggested by theoretical physicist Paul Ehrenfest in 1909.

First, it's important to note that acceleration can involve a
change in speed, a change in direction, or both. Someone on a
merry-go-round rotating at constant speed is accelerating, because
the person is constantly changing direction.

Consider a circular disk at rest. The circumference (C) of the
disk is described by the formula $C = \pi d$, or pi (π, approximately
3.14) multiplied by the diameter (d)—something you probably
learned in high school but may have forgotten. This formula holds
true for the familiar Euclidean geometry implicit in graph paper:
straight lines remain perfectly straight, parallel lines never meet,
and the angles of a triangle always add up to exactly 180 degrees.

In our example, let's say the diameter of the disk is 30 meters.
You might, for example, lay out 30 one-meter sticks end to end,
and that would be the diameter. Similarly, the circumference, which
turns out to be 94.2 meters, could be represented by bending 94.2
one-meter sticks around the rim of the disk.

Now set the disk spinning rapidly. According to special relativity, meter sticks (along with any other objects) shrink in the
direction of motion. The diameter is always at a right angle, or
perpendicular, to the direction of the disk's rotation, so the diameter doesn't shrink. It's still 30 meters. But the meter sticks
around the rim of the disk are moving in the direction of motion,

so they appear to shrink to an observer who is not rotating with the disk. The observer needs extra one-meter sticks, more than 94.2 of them, to span the rim of the disk. The circumference is no longer pi multiplied by the diameter; it's greater than that.

That can happen only if geometry is not flat. It must be distorted or curved, like looking at graph paper through a fisheye lens. And since acceleration is the same as gravity, gravity must curve or distort space-time. Remarkably, Einstein's insight about gravity and acceleration required no advanced mathematics. He somehow got to the heart of the problem without complex equations. But to fully develop the general theory of relativity, Einstein would have to journey into a new, highly mathematical world of curved geometry.

It would take Einstein seven years to complete his masterwork. He would fail, doubt his own intuition, experience bouts of exhaustion, and reach out to a friend in a desperate plea for help. Then at the last minute, a rival making rapid progress threatened to publish the final equations before Einstein did.

DEEPER DIVE: Space and Time, a Perfect Union

More than a century before Einstein, the French mathematician Jean d'Alembert considered time as a fourth dimension. "I said earlier that it is impossible to conceive of more than three dimensions," he wrote in a 1754 encyclopedia article. "A clever acquaintance of mine believes that one might nevertheless consider timespan as a fourth dimension, and that the product with volume would in a certain manner be a product of four dimensions; this idea may be contested but it has, it would seem to me, some merit, if only because of its novelty."

In the 1870s, an American con artist named Henry Slade used the idea of a fourth dimension to his own advantage. Through "automatic writing" allegedly inscribed by ghosts, the extraction of objects from supposedly sealed three-dimensional containers, and the production of mysterious noises, Slade convinced members of London society along with several eminent German scientists—two of them future Nobel laureates—that he had contacted the spirit world through extra dimensions. Although Slade was put on trial in London in 1877 and convicted of fraud (he had already been prosecuted in the United States), the notion of a fourth dimension had made an indelible imprint on the public's imagination.

Could it be that we live in a world with more dimensions than the meager three through which we navigate our daily lives? And if our world does not permit us to see the full nature of the universe, how could an extra dimension ever be visualized?

The answer came not from Einstein but from the mathematician Hermann Minkowski, Einstein's teacher at Zurich Polytechnic. Based on Einstein's work, Minkowski fused the three dimensions of space and one of time into a four-dimensional worldview.

In analyzing Einstein's equations of special relativity, Minkowski realized that relativity could be expressed and understood in terms of geometry. Einstein's equations describe a kind of give-and-take between space and time—as speed increases, the interval between the ticks of a clock appears to lengthen, while moving objects appear to shorten. Two observers moving at different speeds may disagree on the length of time or the distance between two events, but something stays the same for all observers. That something, which Minkowski called the space-time interval, is the four-dimensional analog of what we call length in three dimensions.

It was all oddly reminiscent of a version of Pythagorean's theorem relating the length of a hypotenuse of a triangle to its two sides (see Chapter 2 for further discussion). Just as the hypotenuse, L, can be calculated ($L^2 = x^2 + y^2$) in two dimensions, the space-time interval L in four dimensions has a similar formula—except that time comes in with a minus sign. The four-dimensional L can be calculated as $L^2 = x^2 + y^2 + z^2 - c^2t^2$, where c is the speed of light. The minus sign is critical. Time and space may vary from one observer to the next, but the space-time interval remains the same for everyone.

On a space-time diagram—a graph that plots space against time—it's easy to visualize the give-and-take between space and time. Observer 1 witnesses two events as separated by a shorter distance in space and a longer duration in time than observer 2. But they both agree on the length of the space-time interval.

Think of the interval as the needle of a sundial, suggests astrophysicist Ethan Siegel. As the Sun moves during the day, the shadow of the needle changes in direction and length, but the direction and length of the needle itself remain fixed. Similarly, time and space—the separate, shadow-like components of space-time—vary depending on the motion of an observer, but the space-time interval never changes.

Minkowski understood the deeper implications of this geometry. Crucially, he saw that although the path of an object moving at a constant speed traces out a straight line in space-time, an object undergoing acceleration describes a curved path.

Einstein initially dismissed Minkowski's formulation as "superfluous learnedness." He quipped to a friend: "Since the mathematicians have invaded the relativity theory, I do not understand it myself anymore." But by 1912, Einstein had fully embraced Minkowski's geometric concepts. They were essential for describing the curvature of space-time in Einstein's theory of gravity, and he was the first to admit it. Minkowski did not live to see how his work endured, however. He died of a ruptured appendix at age forty-four in 1909, less than a year after introducing his space-time diagram.

....
....

DEEPER DIVE: Testing the Equivalence Principle before Einstein
Sometime around 1590, Galileo climbed the stone steps to the top of the Leaning Tower of Pisa. With a crowd of students and faculty

assembled below, he dropped two balls of different weight and composition and found that they hit the ground at the same time. At least that's the story that Galileo's assistant, Vincenzo Viviani, told in his biography of the master.

Although Galileo mentioned the idea of dropping balls from a great height in his 1638 treatise *Two New Sciences,* he probably never carried out the experiment. Had he done so, the balls would have fallen so quickly that it would have been difficult to time the event with the clocks Galileo had available. Galileo did perform a gentler, more leisurely experiment in which he rolled balls down a ramp, which reduced the gravitational acceleration and consequently slowed the motion. By timing how long it took for balls of different composition and weight to travel a set distance on the ramp, he showed that all objects accelerate at the same rate, depending only on the ramp's angle of incline and not on the mass of the bodies.

The ramp experiment was one of the first tests of the principle of equivalence, which lies at the heart of Einstein's general theory of relativity. The principle says that over small regions of space, uniform acceleration is the same thing as a uniform gravitational field. That can be true only if two different notions of mass, inertial and gravitational, are identical. Inertial mass refers to the property of matter that resists acceleration (it's why pushing a car is a lot harder than pushing a wheelchair), and it does not depend on gravity. Gravitational mass refers to the property of matter that determines how strongly an object responds to gravity's tug; it's the mass you measure when you step on a scale.

Newton embraced the equality between inertia and gravitational mass, and even began his 1687 masterpiece on the laws of motion, the *Philosophiae Naturalis Principia Mathematica* (*Mathematical Principles of Natural Philosophy*, widely called just the *Principia*), with a statement about their identity: "[Mass] can also be known from a body's weight, for—by making very accurate experiments with pendulums—I have found it to be proportional to the weight." Newton built two identical pendulums by suspending duplicate wooden boxes from separate, eleven-foot-long threads. He treated one box as a reference, keeping it filled with a quantity of wood always weighing the same amount. In the other box, he placed gold of the same weight and set the two pendulums swinging. Newton then repeated the experiment with different materials—silver, lead, glass, salt, water, wood, and wheat—that always weighed the same as the material in the reference box.

Here the inertial mass refers to the mass that must be pushed for the pendulum to swing, and the gravitational mass is the mass that responds to Earth's gravitational pull downward. If the two are equal, then the two pendulums, if released at the same time at the same angle, must keep in step, with the interval between swings dependent only on the length of the pendulum, not on the mass or composition of the material in the box. He found that this equivalence holds to one part in 1,000.

Newton also looked to the heavens for further evidence. He realized that if the equivalence did not hold, the motion of Jupiter's moons about their parent planet would be unstable. If the Sun attracted some of the moons more strongly than others, depending on the amount of their gravitational mass, the Jovian system would collapse. Newton proposed a similar argument for the stability of Earth and its moon. The French mathematician and astronomer Pierre-Simon Laplace refined Newton's argument in 1707, confirming the equivalence principle to a few parts in 10 million.

Testing didn't improve appreciably until the 1890s, when the Hungarian physicist Baron Roland von Eötvös conducted a new kind of experiment. He placed equal weights of different materials on either end of a rod and suspended the dumbbell-like arrangement horizontally from a fine wire. In this setup, known as a torsion balance, Earth's rotation provides a centrifugal force that pushes the inertial mass of each dumbbell outward, away from the other, while Earth's gravity tugs on the gravitational mass. If the two types of masses were unequal, it would cause the horizontal rod to rotate ever so slightly. He and his colleagues found that the inertial and gravitational mass were equivalent up to a few parts in a billion.

2

FROM TURMOIL
TO TRIUMPH

S PACE-TIME IS CURVED. It sags, stretches, and buckles.
Einstein arrived at that stunning realization in 1912, when
he became convinced that his new theory of gravity was not only
geometric but that it would be a radical departure from the flat
realm of points, straight lines, and planes that philosophers, physicists, and mathematicians had used for more than 2,000 years to
describe the natural world.

Einstein had been examining a particular type of uniform
acceleration—a disk rotating at high speed, like a rapidly spinning
merry-go-round. Such an object, which moves at constant speed
but is changing direction, is undergoing uniform acceleration.
Einstein had found that a high-speed merry-go-round creates a
space-time in which all the familiar rules taught in a high school
geometry class—the circumference of a circle is always equal to
its diameter times the number pi, and the shortest distance between two points is a straight line—no longer apply. That can

happen only if space is curved. And by the equivalence principle, if uniform acceleration creates a curved geometry, then so must gravity. In fact, gravity is one and the same as curved space-time.

With that insight, Einstein had an entirely new way of thinking about freely falling objects, which went to the heart of his new theory. Einstein's great insight in 1907, as he was sitting in his chair in the patent office, had been that objects in free fall, like a man falling off a roof, do not feel gravity. If a body in motion feels no force, then it should follow the shortest path possible, a straight-line path. Yet planets in free fall around the Sun move in elliptical orbits, a ball thrown in the air follows a parabolic path as it falls back to Earth, and a beam of light, which always travels the shortest, straightest route between two points, bends in the presence of a massive object.

But Einstein, as if donning special space-time glasses only he knew how to use, looked at those curved paths and saw something different. The paths actually were straight. It was the space-time these freely falling objects traveled through that was curved.

Imagine a caterpillar constrained to move along the surface of a sphere. Determined to follow a straight-line path, the caterpillar never deviates to the right or to the left as it inches forward. Nevertheless, the creature inevitably traces out a circle as it navigates the sphere. Similarly, every object in the universe, including those in free fall, has no choice but to follow the curvature of the local space-time it travels through.

Go back to the caterpillar, which is now no longer on a perfect sphere but on a different kind of terrain. On this surface, as the caterpillar inches along, the traveling suddenly becomes easier. It goes faster with no extra effort. The caterpillar might attribute this to some attractive force that is pulling it forward. The force might even be named gravity. But from our three-dimensional perspective, looking down on the surface, we realize that the caterpillar

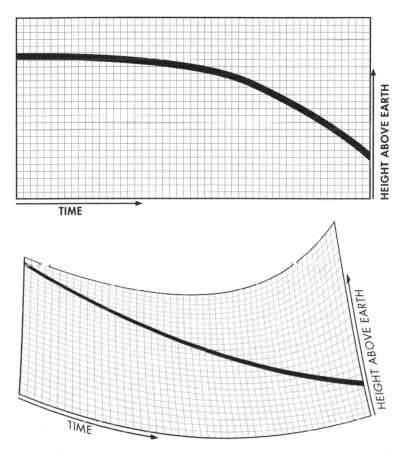

A ball follows a parabolic path as it falls to Earth. According to Einstein, the path appears curved only because Earth curves space-time in its vicinity. If the curvature of space-time is taken into account, the falling ball follows a straight-line trajectory. *(Courtesy Kristen Dill.)*

is simply slipping toward a depression on the surface. What the caterpillar calls gravity we call geometry.

The analogy to our three-dimensional world is that we can't see the curvature of the four-dimensional space-time we inhabit, but we can sense it. We feel the curvature—the local depression in space-time caused by Earth's gravity—every time we roll out of bed in the morning and experience a tug downward.

In place of Newton's idea that a massive body pulls objects toward it because it exerts a gravitational force, the general theory says that a massive body distorts or dimples space-time so that objects fall toward it, like the caterpillar on a downhill slope. Gravity equals curvature. And because Einstein's famous formula, $E = mc^2$, says that mass and energy are two different forms of the same entity, both can generate curvature. Einstein's description of gravity wasn't just different from Newton's; it was more accurate. With one notable exception, Newton's law of gravitation works beautifully in the relatively weak field of the solar system. But it fails to describe the motion of stars zipping around a black hole or another extremely dense, massive body. For an accurate description of those extreme cases of gravity, the general theory of relativity was a necessity.

Einstein's theory also explains a long-standing puzzle about the motion of Mercury. The innermost of the solar system's planets moves about the Sun in an elliptical path. Actually, it's not *perfectly* elliptical, because each time Mercury completes one orbit, the planet begins the next orbit at a point ever so slightly ahead of where it began its last. As a result, the point at which the planet lies nearest to the Sun, known as the perihelion, slowly rotates about the Sun, a motion known as precession. Although the orbits of all the planets precess, it's only Mercury's whose deviation from Newtonian gravity is large enough to be easily detected. As viewed from Earth, Mercury precesses by 5,600 arcseconds every century (1 arcsecond is 1/3,600 of a degree). Newton's law predicts a precession of 5,557 arcseconds per century, due to a nonrelativistic effect—the tug of all the other planets. The remaining 43 arcseconds could not be explained, and some astronomers had even invoked the existence of an unknown planet, dubbed Vulcan, to account for the discrepancy. A search for the planet had come up empty-handed. A hunt for a putative group of asteroids or field of dust near Mercury that scientists suggested might account for the discrepancy also came to naught. But instead of searching for objects in space, scientists should have examined the nature of space itself. The amount of space-time curvature around the massive Sun predicted by Einstein's complete 1915 theory neatly explained the full precession.

General relativity also involves much more than substituting curved geometry for Newton's gravitational force. In Newtonian gravity, space and time are the featureless backdrops, the silent and immutable stage upon which the universe's actors—humans, bowling balls, planets, and stars—strut their stuff. Even in Einstein's special theory of relativity, which wove space and time into a single fabric, the ticks of a clock and the markings of a ruler are spectators, exerting no influence on the comings and goings of the cosmic players.

Einstein's general theory, however, demands that the stage is an equal partner in the action, a malleable and dynamic participant. When a heavy object makes its entrance, the space-time stage sags and stretches. In turn, those contortions push and pull the players, on occasion even trapping them in giant sinkholes (black holes) from which they never emerge. As the players move, redistributing their mass and energy, they change the shape of the stage. In this never-ending cosmic dance, as the theoretical physicist John Archibald Wheeler would later put it, mass tells space-time how to curve, space-time tells matter how to move.

But therein lay a challenge for Einstein. He had to express the intimate link between matter and space-time in a mathematical language. Because gravity manifested itself as curvature, the effort would require Einstein to become an expert in differential geometry, a way of measuring and characterizing the twists and turns of a curved surface. Einstein's insistence that physical laws, including gravity, look the same to all observers, regardless of their motion, imposed further requirements on the mathematics. The equations would have to be generally covariant—that is, they would have to take the same form regardless of the coordinate system one might use to measure gravity (curvature) or even just to describe the positions of objects. For instance, the equations should look the same and encode the same information whether acceleration is measured in units of meters and seconds or units of miles and hours, or whether an observer's frame of reference has undergone some arbitrary rotation or other change.

For Einstein, who had never had much use for fancy mathematics and had thought it could even obscure the underlying physics, the task was onerous. He would later recall that in 1913, Max Planck, the veteran German physicist who was one of the originators of quantum theory, warned Einstein of the challenge that lay ahead in developing his theory. "As an older friend I must advise you against it," he said "for in the first place you will not succeed, and even if you succeed no one will believe you."

Knowing he was in over his head, Einstein turned to his college classmate Marcel Grossmann with a plea for help. Grossmann, who had become a mathematics professor specializing in geometry at their alma mater, was happy to oblige.

It was not the first time he had come to Einstein's rescue. The two had first met as students at the Swiss Federal Polytechnic School in Zurich, where Einstein studied physics and mathematics from 1896 to 1900 along with his future wife Mileva Marić. While Einstein skipped classes (especially mathematics) that didn't interest him and developed a reputation as a rebel who resisted instruction and alienated his teachers, Grossmann was organized, well liked, and studious. His carefully annotated notebooks on class lectures proved a lifeline for Einstein, keeping him on the path to graduation.

Even so, Einstein's strained relationships with his teachers made it impossible to find a university job after graduation. He had taken for granted that he would become an assistant to one of his professors at Polytechnic. Among the five in his graduating class, only Einstein did not receive an offer.

In vain, he contacted well-known scientists around Europe to seek employment, growing more desperate for permanent work after May 1901, when Marić, his girlfriend, became pregnant.

In one letter, sent to Wilhelm Ostwald, a noted chemistry professor at the University of Leipzig who would go on to win a Nobel Prize, Einstein's job inquiry became a plea: "I am without money, and only a position of this kind would enable me to continue my studies."

When a follow up letter got no response, Einstein's father, in poor health and a stranger to the world of academia, sent his own note: "Please forgive a father who is so bold as to turn to you, esteemed Herr Professor, in the interest of his son. Albert is 22 years old, he studied at the Zurich Polytechnic for four years, and he passed his exam with flying colors last summer. Since then he has been trying unsuccessfully to get a position as a teaching assistant.... I can assure you that he is extraordinarily studious and diligent and clings with great love to his science. He therefore feels profoundly unhappy about his current lack of a job, and he becomes more and more convinced that he has gone off the tracks with his career."

Einstein never did get an offer from Ostwald. But Grossmann, whom Einstein had confided in about his plight, came through for his friend. Grossmann's father was an old friend of the head of the Swiss patent office in Bern, a connection that secured Einstein a job as patent examiner second class in 1902 and enabled him to marry Marić soon after.

In the meantime, Grossmann's career had flourished. The same year Einstein began working at the patent office, Grossmann earned his doctorate, and five years later he became a full professor of descriptive geometry at the Polytechnic in Zurich, which had been renamed the Eidgenössische Technische Hochschule, or ETH. During the winter term of 1911/1912, Grossmann became chair of the department for mathematics and physics teachers.

Einstein remained at the patent office until 1909, when he got his first full-time academic position as an associate professor of physics at the University of Zurich. In 1911, the German University in Prague lured him from Zurich with an offer of a full professorship, but Einstein stayed for only sixteen months. In 1912, his reputation on the rise, Einstein received several offers for faculty positions, including one at ETH. The very institution that had denied Einstein a position as an assistant to a professor now wanted him back as a full professor. Grossmann, of course, helped conduct the negotiations that brought him there.

Einstein arrived at the ETH in August 1912. That was when he so urgently sought Grossman's help.

In looking over his own university notes, Einstein realized the revolution had started without him. Six decades earlier, two brilliant scholars at the University of Göttingen, Carl Friedrich Gauss, dubbed the prince of mathematics, and his student, Bernhard Riemann, had explored the geometry of curved surfaces.

For more than 2,000 years, traditional geometry, in which parallel lines never meet, space consists of three dimensions, and the angles of a triangle always add up to 180 degrees, had worked perfectly.

In the fifth century BCE, Pythagoras and his protégés, working in southern Italy, developed a set of mathematical rules based on the flat, two-dimensional geometry of the ancient Egyptians and Babylonians. The famous Pythagorean theorem, $a^2 + b^2 = c^2$, refers to the three sides of a right triangle, in which the two legs, a and b, meet at a 90-degree angle and c is the diagonal, or hypotenuse. Not only is c the longest side of a right triangle, but it is also the shortest distance between the two points located at either end of c.

Some 200 years after Pythagoras, Euclid of Alexandria focused on the same kind of geometry—points and lines confined to a plane. In book one of his thirteen-volume treatise *The Elements*, Euclid presented five postulates that for two millennia defined geometry and were held as absolute truths. The postulates seemed irrefutable, in part because they appeared obvious. The first four postulates certainly fit that description: a straight line always passes between two points; a line segment can be indefinitely extended in either direction; given any straight line segment, a circle can always be drawn such that the segment is the radius of the circle; all right angles are equal to each other.

The fifth postulate was more complex. Restated in modern terms, it says that given a straight line and a point not on that line, only one line can be drawn through that point that is parallel to the original line. The postulate is exactly equivalent to two other formulations: that the three interior angles of a triangle always add up to 180 degrees and that the ratio of the circumference of a circle to its diameter is always the number pi.

For centuries, mathematicians tried to prove that the fifth postulate was a direct consequence of the first four, but all attempts failed. Still, Euclidean geometry reigned supreme. It wasn't just a tool for analyzing nature; nature seemed to live by its rules. Palaces and churches embodied Euclid's straight lines and angles; the geometry had permeated every facet of civilization.

Then, in the early 1800s, mathematicians discovered two new geometries, both curved, that not only refuted the fifth postulate but also allowed them to think in new ways about the shape and nature of the universe. Mathematicians, if not physicists, were no longer imprisoned by the straight line; space was no longer parceled only into the rectilinear grids of graph paper. Four mathematicians played a key role in this upheaval: Gauss, Riemann, János Bolyai, and Nikolai Lobachevsky.

Bolyai grew up in the mountains of Transylvania, far from the centers of mathematics in France, Germany, and England. But as a young man he became obsessed with the fifth postulate. His father, Farkas Bolyai, may have sparked his interest. Farkas, who had studied under Gauss at Göttingen, had twice believed he had found a proof that the fifth postulate could be derived from the other four, only to be shot down each time by Gauss. Knowing all too well his own exhaustion in trying to prove the theorem, he warned his son that the study would rob him of health, peace of mind, and happiness. "I know this way to the very end," he wrote his son in 1820. "I have traversed this bottomless night, which extinguished all light and joy in my life. I entreat you, leave the science of parallels alone. . . . Learn from my example." The great Gauss, who was often taciturn, refused to take Bolyai on as a student, but the young mathematician persevered.

Bolyai at first set out to prove the fifth postulate by an oft-used strategy in science, reductio ad absurdum. In this approach, a researcher assumes a postulate to be false and then shows that such an assumption leads to contradictory conclusions, thus demonstrating that the original postulate must in fact be true. Instead, in the 1820s, he found an "imaginary geometry," curved like a Pringles potato chip. In this geometry, called "hyperbolic" because of its saddle-shaped appearance, the sides of a triangle added up to less than 180 degrees and there existed an infinite number of lines parallel to a given line. He had, he declared, discovered "such wonderful things that I was amazed": a mathematical paradise, a world of curved geometries that was very different from the flat space of Euclid.

But when Bolyai contacted Gauss, the venerable mathematician refused to praise him, asserting that he himself had found, though never published, a similar geometry years earlier. Indeed, Gauss often did not publish work he deemed controversial. He would not publicly go against the prevailing thinking of mathematicians or philosophers such as Immanuel Kant, who asserted that Euclidean geometry was an intuitive, universal truth.

Yet letters to friends reveal Gauss's thinking. In 1824, he wrote "The assumption that [in a triangle] the sum of the three angles is smaller than 180° leads to a geometry that is quite different from ours (Euclidean), which is consistent, and which I have developed quite satisfactorily."

Meanwhile, Bolyai published his work in *The Absolutely True Science of Space,* part of a larger treatise written by his father and published in 1831. It received scant attention. Already disheartened by Gauss's reaction, Bolyai would later discover that another mathematician had already published similar work.

In the Russian city of Kazan, mathematician Nikolai Ivanovich Lobachevsky had found the same type of "imaginary geometry"—hyperbolic geometry. His article on the finding, published in 1829–1830 in the *Kazan Messenger,* a publication of the University of Kazan written in Russian, was not widely read.

Both Bolyai and Lobachevsky died without knowing their work would have a lasting impact. A discouraged Bolyai became a recluse, leaving behind 20,000 pages of mathematical manuscripts at his death at age fifty-seven in 1860. A few years later, Lobachevsky also died in obscurity, nearly blind and unable to walk. It would be up to Gauss and his student Riemann to continue the work that Einstein would so radically embrace in his general theory of relativity.

While managing a fraught personal life—caring for his sick mother and constantly arguing with his wife, who wanted to move from Göttingen to Berlin—Gauss developed an interest in differential geometry. He realized that the relationship between lengths spelled out by the Pythagorean theorem, true for flat surfaces, could be modified to measure the intrinsic curvature of a surface. He also demonstrated that a single number, the Gaussian curvature, could fully describe the curvature of a surface, such as that of a cylinder or sphere.

It was Gauss's student Riemann who made the leap to higher dimensions. The shy and sickly second child of six born to a poor Lutheran minister and his wife in Breselenz, part of the Kingdom of Hanover, Riemann early on amazed family and teachers with his mathematical abilities. Initially his father tutored him, but at age ten Riemann began instruction with a professional teacher who found that his student's proofs were sometimes better than his own.

When his family scraped up enough money to send him to the University of Göttingen in 1846, Riemann intended to become a priest like his father. But he supplemented his studies in theology and philosophy with mathematics, including classes taught by Gauss. Ultimately, his father gave Riemann permission to pursue a degree in math. After additional study at the University of Berlin, Riemann returned to Göttingen in 1849. There his interest in geometry blossomed. For his doctoral thesis, supervised by Gauss and completed in 1851, Riemann showed that a set of exotic numbers, those with a component proportional to the imaginary number square root of –1, could be expressed as a curved surface.

Next, he worked on a separate degree that would qualify him to be a lecturer at a university in Germany and support himself. To earn that degree, Riemann had to give a *Habilitationsvortrag*, a probationary lecture. Like all students, he presented three topics to his professors; the first two he had thoroughly researched, but the last one, on the foundations of geometry, he had not prepared as extensively. The faculty traditionally chose one of the first two topics on a student's list, but Gauss, with his interest in geometry, could not resist selecting Riemann's third, less well-researched choice.

In anxiously preparing for the geometry talk and fighting a life-long fear of public speaking, Riemann fell ill and had to cancel his scheduled presentation date. He had to cancel a second time when Gauss, then in his seventies, became ill. Riemann finally delivered his talk, "On the Hypotheses That Lie at the Foundation of Geometry," on June 10, 1854. Gauss may have been the only one in the audience to fully comprehend the implications of Riemann's talk, but experts have since hailed the lecture as one of the most farsighted in the history of mathematics. Riemann was only twenty-seven.

In his talk Riemann extended Gauss's work on measuring curvature in two dimensions so that it could be applied to higher dimensions—three, four, or any number. Riemann also considered surfaces where the curvature might vary not only from point to point but also along different directions from the same point. The thinking was anathema to some.

Einstein needed Grossmann's help in navigating the new and difficult mathematics of Riemannian curvature.

"Grossmann," shouted Einstein as he entered his friend's home, "you've got to help me or else I will go crazy!"

Because a curved surface can vary in a complicated way, Grossmann had to teach Einstein about tensors, mathematical objects that keep track of more than one variable at a time. In particular, Grossmann introduced Einstein to the Riemann tensor, which directly measures the curvature of space. Einstein embraced the Riemann tensor, adapting it to describe the curvature not just of space, but of space-time. The tensor provided Einstein with a way of equating sources of gravity—mass, momentum, and energy—with space-time curvature in a form that would look the same in all coordinate systems (see "Deeper Dive: Riemann's Work and the Metric Tensor").

Grossmann also introduced Einstein to differential calculus, a way of calculating quantities on surfaces of arbitrary curvature, started by Elwin Bruno Christoffel and fully developed at the University of Padova by Gregorio Ricci-Curbastro and his student Tullio Levi-Civita.

One thing was certain. Einstein had never worked so hard in his life. Compared to the problem of gravity, he wrote to a friend in 1912, the theory of special relativity was "mere child's play." He also developed a new respect for mathematics.

Einstein and Grossmann agreed to work together. They wrote many of their calculations in a small brown notebook, the Zurich notebook, that documented their year-long collaboration.

In 1913 they jointly published a paper, "Entwurf einer verallgemeinerten Relativitätstheorie und einer Theorie der Gravitation" (Outline of a generalized theory of relativity and a theory of gravitation), commonly referred to as "Entwurf." Einstein wrote the section dealing with physics, and Grossmann wrote the mathematical part. The equations developed by Einstein and Grossmann in "Entwurf" closely resembled those that would become the complete general theory of relativity. The mathematics was indeed generally covariant, describing the curvature due to gravity in the same way in all frames of reference. Success was at hand.

But then Einstein and Grossmann backed away from what they had accomplished. They found that their equations did not reduce to Newton's law of gravity, as they must, when gravity is weak and does not vary with time. Just as worrisome, the formulation did not seem to conserve energy.

The only way around those apparently fatal flaws, they believed, was to abandon the idea of general covariance. The new equations allowed a description of gravitational curvature that had the same form for some observers in relative motion, but not all. Einstein even devised a proof that seemed to show that general covariance was impossible. Only limited covariance seemed to be permitted.

At first Einstein boasted to friends that his theory of gravity was nearly complete. But he eventually grew worried that giving up on general covariance would mean giving up on the principle of equivalence: uniform acceleration could not *always* be replaced by a uniform gravitational field, it seemed. The equivalence principle had been Einstein's North Star, guiding him since 1907 to understand the nature of gravity and space-time. In his heart, Einstein never abandoned the principle. It was simply on hold.

In the meantime, Einstein had moved from Zurich to Berlin. Three prominent German scientists, the chemist Fritz Haber and the physicists Planck and Walther Nernst, had lured Einstein with an offer he could not refuse: membership in the Prussian Academy of Sciences, a position as professor without any teaching responsibilities, and a promise to be appointed director of a planned Kaiser Wilhelm Institute for theoretical physics. With the growing recognition of Einstein's achievements, the powers that be in Berlin overlooked that he had renounced his German citizenship, never served in the Prussian army, and, in a time of blatant anti-Semitism, was Jewish.

But there was a more personal reason Einstein wanted to relocate to Berlin. Relations between Einstein and his wife had been strained for several years. The two had been inseparable during their university years at Polytechnic. After they married, in the evenings when Einstein came home from the patent office, they would work together on physics problems by the light of a kerosene lantern. But Marić, having twice failed the final oral examination at Polytechnic, never graduated and felt increasingly left out of her husband's academic life and ignored at home. For his part, Einstein, who had affectionately nicknamed Marić his "dolly" during their courtship, became distant and cruel.

During a visit to Berlin in 1912 to visit his widowed mother, Einstein renewed his friendship with a cousin, Elsa Einstein

Lowenthal, and they soon began an affair. He told Elsa that he loved her and not to worry about Marić. "I treat my wife as an employee I can not fire," he wrote.

Marić and their two young children moved to Berlin shortly after Einstein's arrival there in April 1914. They didn't stay long. Three months after they arrived, she and the children went back to Zurich for good.

The year proved tumultuous for Einstein in other ways. On August 1, Germany declared war on France and England. Although Einstein had years earlier renounced his German citizenship and had become a citizen of neutral Switzerland, he was now working behind enemy lines. All communication was severed with scientists from other parts of western Europe. In Berlin, British and French nationals, or people suspected of being citizens of those countries, were chased and attacked. Either out of patriotism or from fear of loss of sales, owners of restaurants with English-sounding names renamed their shops; the popular Café Windsor became Kaffee Winzer. When war broke out, the German astronomer Erwin Finlay Freundlich had just arrived in Crimea to study the solar eclipse of August 21, 1914. He had mounted the expedition to test Einstein's theory of gravitational light bending. But as soon as war was declared, he and his associates were promptly interned in Russia as enemy aliens, their equipment confiscated.

In October 1914, ninety-three German scientists signed a proclamation giving their unqualified support to the German military. Einstein refused to sign the "Manifesto of the Ninety-Three" and instead was one of just four scientists to endorse a proclamation protesting Germany's aggression.

Using his extraordinary powers of concentration, Einstein blocked out the war as best he could. But sometimes it was too close to be ignored. Einstein's office was in the Kaiser Wilhelm Institute for Physical Chemistry. The institute's director was Fritz

Haber, who had befriended Einstein when he came to Berlin. To aid in the war effort, Haber and his collaborators had begun experiments with chlorine, a poisonous gas. On December 17, 1914, a test tube of cacodyl chloride, an unstable substance, caught fire in Haber's laboratory. The subsequent explosion blew off the right hand of one researcher and killed another. Fortunately for Einstein, his office was unharmed.

Einstein did his best to ignore not only the war but some of the criticism of the "Entwurf" paper he had written with Grossmann. Another college classmate and friend, Michael Besso, had warned Einstein that the "Entwurf" equations would not allow the uniform acceleration of someone on a merry-go-round to be interpreted as a gravitational field, another apparent blow to the equivalence principle. Einstein paid little attention. The 1913 theory also did not account for the precession of Mercury's orbit about the Sun. Yet at the end of 1914 Einstein was enamored enough with the theory to publish a long explanatory paper.

For the first part of 1915 Einstein worked on other topics, even dabbling in laboratory work on magnetism. In late June, he traveled to Göttingen, where the mathematician David Hilbert, the heir apparent to Gauss, had invited him to give a series of talks on relativity.

Afterward, Einstein wrote to his friend Arnold Sommerfeld that he was enthusiastic about meeting Hilbert and had great joy in convincing him (and another Göttingen mathematician, Felix Klein) of the validity of his theory. Indeed, Hilbert was so taken with Einstein's presentations that he began studying the mathematics of relativity.

In September 1915, Einstein took a break, vacationing in Switzerland and visiting his family. Soon after he returned, in October, Einstein was ready to admit that the 1913 theory he had collaborated on with Grossmann was wrong. The principle of equivalence was too important to ignore.

Desperate to save his theory, he let mathematics be his guide. Returning to the covariant equations he and Grossmann had developed but decided must be wrong, Einstein made the happy discovery that he had been mistaken. By carefully going through the mathematics, he found that the equations *did* conserve energy. They *did* reduce to Newton's law of gravity in the weak gravitational field limit.

Sometime around late October, Einstein received a disturbing letter from his friend Sommerfeld. He was not the only one working to revise his theory, Sommerfeld wrote. Hilbert also believed the "Entwurf" work was flawed and was formulating his own version.

After working on the general theory for eight years, Einstein would not be usurped. In a frenzy of activity in November 2015, Einstein gave four talks, one every Thursday, to the Prussian Academy of Sciences. Notably, the talks did not build on one another but instead revealed the zigzag of creative thought as Einstein grappled with tensor mathematics and the physics. Punctuating the lectures, he and Hilbert exchanged a flurry of postcards about each other's progress—and the implicit race to the finish line. In one of the missives, Einstein made clear that Hilbert's system of equations was identical to what he himself had found several weeks earlier and had known about since working with Grossmann in 1912.

In Einstein's first talk, on November 4, he presented new equations to describe gravity, renouncing his "Entwurf" work and a subsequent 1914 paper. But within days, he realized the new work was not correct. The second talk, on November 11, rejected the first presentation and introduced a new equation. On November 18, a jubilant Einstein revealed exciting news: he had derived the proper advance of Mercury's orbit from his new equation.

On November 25, his last presentation, Einstein rejected the equations from the first two talks. In one of the most remarkable discoveries of the twentieth century, he finally unveiled the right solution to describe the inviolable link between space-time and gravity / matter-energy.

Einstein's compact equation takes up but a single line:

$$R_{\mu v} - 1/2R \, g_{\mu v} = 8\pi G_N/c^4 \, T_{\mu v}$$

Yet this equation has kept on giving for more than a century (see "Deeper Dive: The Meaning of Einstein's Equation"). It has revealed that the universe is expanding, that spinning objects drag space-time along with them the way the blades of a blender drag pancake batter, and that gravity acts as a zoom lens to reveal some of the first galaxies born in the universe, nearly 14 billion years ago. The equation also revealed the existence of that most phantasmagorical of all objects, a black hole.

Back in 1915, with the war raging in Europe and a British blockade in full swing, few people outside of Germany knew of Einstein's work. But Einstein was able to send a copy of his work to his colleague Willem de Sitter in neutral Holland. He in turn sent a copy to British astronomer Arthur Eddington, who would play a starring role in confirming the general theory, capturing the attention of a worldwide audience and turning Einstein into the first celebrity scientist.

....

DEEPER DIVE: Riemann's Work and the Metric Tensor

Riemann's description of curvature was an essential ingredient of the mathematics Einstein used to develop his general theory of relativity. He could not have arrived at the final form of his theory without it.

To illustrate what Riemann did, recall that in two dimensions, the Pythagorean theorem states that $l^2 = (x_1 - x_2)^2 + (y_1 - y_2)^2$ where (x_1, y_1) and (x_2, y_2) are two points on a flat piece of graph paper and l is the distance between them. Bring those two points close together so that they are separated by an infinitesimal difference along the x axis (call that tiny distance dx) and an infinitesimal distance along the y axis (call it dy). Then the infinitesimal length dl would obey the formula $dl^2 = dx^2 + dy^2$. This is just the Pythagorean theorem applied to tiny distances.

Now let's branch out and go to four dimensions—three of space (length, width, and height) and one of time. Consider two events, separated by an infinitesimal amount of time, dt, and an infinitesimal separation in width, length, and height of dx, dy, and dz. If space-time is perfectly flat, then the interval between these two events can be calculated from the Pythagorean theorem extended to four dimensions: $dl^2 = dx^2 + dy^2 + dz^2 - dt^2$. (The minus sign in front of the time coordinate encodes that the speed of light is a constant.) If we wanted to get fancy, we could write this very same expression as $dl^2 = g_{xx}dx^2 + g_{yy}dy^2 + g_{zz}dz^2 + g_{tt}dt^2$, where the g coefficients in front of the spatial coordinates are set equal to 1 and the g coefficient in front of the time coordinate is −1.

In curved space-time, the interval dl is a more complicated sum. Computing the infinitesimal length now involves the product, two at a time, of every possible grouping of coordinates—the usual x^2, y^2, z^2, and t^2, but also xy, xz, xt, yz, and yt. Moreover, the g terms in front of each of these products are no longer equal to 1. In fact, they may vary according to the location in space and time. So we have

$$dl^2 = g_{xx}dx^2 + g_{yy}dy^2 + g_{zz}dz^2 + g_{tt}dt^2 + g_{xy}dxdy + g_{xz}dxdz$$
$$+ g_{xt}dxdt + g_{yz}dxdy + g_{yt}dydt + g_{zt}dzdt.$$

The g coefficients, collectively known as the metric tensor, completely describe the curvature of a four-dimensional space-time. What's more, although the value of each g coefficient changes depending on the coordinate system someone might choose to measure the curvature, the form of the equation and the actual infinitesimal length, dl, always stay the same.

Einstein relied on the metric tensor to describe how mass and energy curve space-time.

DEEPER DIVE: The Meaning of Einstein's Equation

Einstein's formula $R_{\mu\nu} - 1/2R \, g_{\mu\nu} = 8\pi G_N/c^4 \, T_{\mu\nu}$ may look mysterious, but its essence can be understood in a simple but profound statement about matter and space-time. To see this, consider each side of Einstein's equation separately.

The left-hand side describes the curvature of space-time. It includes $g_{\mu\nu}$, the metric tensor developed by Riemann (see "Deeper Dive: Riemann's Work and the Metric Tensor"). Curvature in Einstein's theory replaces the concept of a gravitational force in Newton's work.

The right-hand side, which features an object known as the stress-energy tensor, $T_{\mu\nu}$, represents all properties of a material that can curve space-time—not just mass but energy (recall that Einstein said that mass and energy were equivalent), momentum, and pressure.

So we can symbolically write the equation in a much simpler form, suggests astrophysicist and author Jean-Pierre Luminet:

$$\textbf{\textit{G}} = \textbf{\textit{T}}$$

Here, $\textbf{\textit{G}}$ stands for the geometry of space-time and $\textbf{\textit{T}}$ stands for matter. Geometry equals matter. That's what it boils down to. The boldface indicates that $\textbf{\textit{G}}$ and $\textbf{\textit{T}}$ are not mere numbers. They are tensors, because they keep track of more than one direction or variable. Moreover, in general relativity, each tensor has ten independent components, so the single formula represents ten equations.

3

EDDINGTON ON A MISSION

H E WAS OBSESSED with numbers at an early age, learning the 24 × 24 multiplication table before he could read and asking to be taken to the promenade of his seaside English town so he could count the stars. Reserved, studious, and a devout Quaker like his parents, he won a scholarship to Owens College in Manchester when he was only fifteen. By the summer of 1918, the British astronomer Arthur Eddington had become an expert on the structure of stars, mastered mathematical physics, and been appointed director of the Cambridge Observatory.

He was also in danger of going to jail.

Eddington's troubles had begun two years earlier. By March 1916, World War I had raged for nineteen months, and the all-volunteer British army was running out of human fodder. Poison gas and machine guns had killed hundreds of thousands of soldiers in a no-man's-land of barbed wire and trenches. The government decided to make military service compulsory.

As a lifelong member of the Quaker faith's Society of Friends, Eddington was a pacifist who had been ready to request an exemption from service on religious grounds. But senior members of the Cambridge astronomy department quickly intervened to ensure that he would be excused based solely on the importance of his research. The department dared not risk the humiliation and public ire of having a "conchi"—a conscientious objector—in their midst. The press called them cowards and degenerates.

Conscientious objectors were sent to work camps or, if their exemption was denied, were jailed. One Quaker told of being taken to an isolated area and beaten until he could no longer stand, then force-fed when he still refused to obey the command of a colonel.

The war was ever-present. Even on the historic Cambridge campus, Eddington could not escape it. Not far from the portico of Cambridge Observatory, where he had taken up residence with his mother and sister in 1914, Eddington could see courtyards that had been converted into army training grounds and a study hall that had become a soldiers' mess hall. Five undergraduates and fifteen graduate students were arrested for refusing military service.

Even so, Eddington refused to keep a low profile. At a time when a Cambridge professor proclaimed that "the Germans are congenitally unfit to read our poetry; the very structure of their organs forbids it," and the prestigious British journal *Nature* published articles decrying the inferiority of German science, Eddington publicly urged English astronomers to keep the wartime horrors separate from their work and remain collegial with their German counterparts.

Eddington himself had become enamored of a radical new theory of gravity proposed by German-born physicist Albert Einstein. Direct communication with Einstein, who lived in Berlin, was impossible. But Einstein was still able to visit colleagues in the Netherlands, which had stayed neutral. There he tutored astronomers such as Willem de Sitter about his new theory. It was de Sitter who smuggled copies of Einstein's seminal 1916 papers on the general theory of relativity to Eddington. They may have been the only copies available in the United Kingdom until the war ended.

In those papers, Einstein replaced Isaac Newton's view of gravity as a force that acts across space with the idea that gravity *is* space. Specifically, he said, space and time, instead of being stiff and unchanging, are as jiggly as Jell-O. A massive body warps or curves this wobbly space-time much the way a heavy sleeper sags a mattress. A marble rolls toward a heavy body not because of a force but because the heavy body has dimpled the space-time through which the marble must travel.

Eddington didn't just embrace the theory; he became its chief advocate. He exhorted his colleagues at scientific conferences to embrace the theory, published review articles on Einstein's mysterious concept of curved space-time, and defended the work when critics tried to disparage it. Eddington was also one of the few scientists—some said he was the only person in England—to comprehend the theory's complex mathematical language.

Eddington was keen to test the theory, and as early as 1911, Einstein had suggested a way to do so. If a body is massive enough— like the Sun—then it should be possible to observe the curved or bent path of all objects traveling in its vicinity, even particles of starlight. The bending of starlight would show up as a change in the apparent position of the star compared to its position when the Sun was in another part of the sky.

Under ordinary conditions, attempting such an observation would be folly. The blinding light of the solar disk would completely swamp the much fainter light from surrounding stars. But the stars pop into view during those rare times and places when the clockwork motion of the solar system places the Moon directly between the Sun and Earth. Because of the amazing cosmic

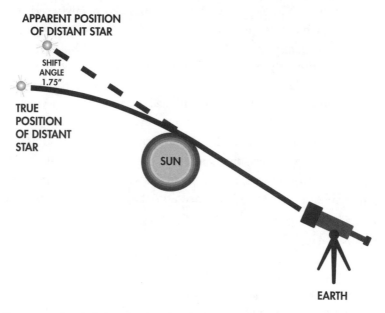

Illustration of starlight bent by the Sun. The bent path shifts the apparent position of a star by 1.75 arcseconds, as predicted by Einstein's theory. *(Courtesy Kristen Dill.)*

coincidence that the Moon is 1/400th the size of the Sun yet is 400 times closer, the Moon blots out the entire Sun, creating a total solar eclipse.

The first eclipse expeditions to test Einstein's theory did not go well. Rain doomed an expedition by Eddington and a colleague to photograph the starlight-bending effect during the total solar eclipse of October 10, 1912, from Brazil. Early in August 1914, an expedition led by astronomer Erwin Finlay-Freundlich and his colleagues left Germany to observe the August 21, 1914, eclipse from Russia. By the time the German scientists arrived in Crimea, World War I had just begun. Finlay-Freundlich and two colleagues were promptly arrested as spies by local authorities, their telescopes confiscated as enemy surveillance equipment. The astronomers were freed just a few weeks later in a prisoner exchange, but their equipment wasn't returned for several years.

The foiled observations only gave the search added allure. It also proved lucky for Einstein. Late in 1915 he realized that the correct deflection of starlight was actually twice what he had calculated in 1911, before he had fully developed his new theory of gravity. Had earlier observations recorded light bending, they would not have found the (incorrect) deflection Einstein had originally predicted, and thus his theory might (temporarily) have been relegated to the garbage heap.

Einstein's revised calculation also sharpened the distinction between his work and that of scientists who adhered to the Newtonian view of gravity as a force of attraction between objects. His new number was twice what scientists had calculated based on Newton's laws. The search for the bending of starlight could now be reframed as a fundamental question: Who was right about gravity and the nature of the cosmos—Newton, whose laws of motion had successfully explained nearly every aspect of the physical world for more than two centuries, or the upstart Einstein, with his radical notion of space-time?

William Campbell, a pioneer in solar eclipse photography from Lick Observatory in California, had hoped to find an answer by observing an eclipse on June 8, 1918, but clouds obscured the stars he needed to detect. More than a year before that attempt, Eddington and Britain's Astronomer Royal, Frank Dyson, had already begun looking ahead to a different eclipse, one that would occur on May 29, 1919.

The 1919 event had several things going for it, Dyson pointed out to readers of *The Observatory,* the Royal Astronomical Society's monthly newsletter. The eclipse would last more than six minutes, one of the longest in the twentieth century. What's more, the occluded Sun would pass through a rich background of stars, the Hyades cluster, providing a bounty of stars with which to test

Einstein's light-bending prediction. Another plus: these stars were relatively bright. That would be important, noted Dyson, because during an eclipse, the sky isn't entirely dark. The Sun's hot outer atmosphere, or corona, which is normally invisible, glows like a halo around the occluded Sun, making it difficult to image dim stars in the vicinity.

While Dyson and Eddington contemplated plans to observe the 1919 event, which would require travel to remote parts of the world, the war raged on. By early 1918, the British military was desperate to replace the hundreds of thousands who had died. Exemption cases came under harsher scrutiny. In January, a military tribunal argued that Eddington, single and thirty-five years old, should have his occupation-based exemption terminated in three months. In April, he was granted a three-month extension, but at a hearing on June 14, 1918, in Cambridge, the military succeeded in revoking Eddington's exempt status.

According to the *Cambridge Daily News,* the chair of the committee, a Major S. G. Howard, "suggested that Prof. Eddington's ability be better employed in active prosecution of the war if placed at the disposal of the Government." In reply, Eddington stood and declared, "I am a conscientious objector."

But the tribunal refused to consider his claim. A man could be exempt based on his work or his religion, but not both. Although Eddington explained that he had initially requested conscientious objector status, "that question is not before us," the tribunal concluded.

Eddington pleaded his case before a local tribunal in Cambridge on June 28, asking if it would consider his application as a conscientious objector. "My objection to war is based on religious grounds," he told the tribunal. "I cannot believe that God is calling me to go out and slaughter men, many of whom are animated by the same motives of patriotism and supposed religious duty that have sent my countrymen into the field." He went on more bluntly: "Even if the abstention of conscientious objectors were to make the difference between victory and defeat, we cannot truly benefit the nation by willful disobedience to the divine will."

Upon deliberation, the tribunal considered the case "a very hard one—hard against Prof. Eddington," the *Cambridge Daily News* reported. But instead of making a final decision, the tribunal gave Eddington until July 11 to get permission from the Ministry of National Service for the court to consider his claim.

In the meantime, Cambridge astronomers tried one more time to save Eddington, requesting an exemption based solely on his astronomical studies. The effort appeared promising; all Eddington had to do was to sign and return a letter that noted his work was of national importance. He did so but felt compelled to add a postscript. Eddington wrote that even if his work was deemed not to be of national importance, he would ask to be excused as a conscientious objector. His note defeated the letter's intent. Eddington's Cambridge colleagues were furious.

The ministry did, however, give its approval for the tribunal to consider Eddington's status as a conscientious objector when it reconvened on July 11. By then, Eddington was desperate. He solicited letters of support from colleagues, only some of whom responded. At the hearing, Eddington presented a letter from Dyson, who, in addition to being Astronomer Royal, had become chair of the Joint Permanent Eclipse Committee of the Royal Society and the Royal Astronomical Society.

Dyson carefully composed his letter to appeal to the sensibilities of the tribunal. After extolling Eddington's work, Dyson noted that the astronomer's studies "maintain the high tradition of British science at a time when it is very desirable that it should be upheld, particularly in view of a widely spread but erroneous notion that the most important scientific researches are carried out in Germany."

Then Dyson came to the punch line: "There is another point to which I should like to draw attention. The Joint Permanent Eclipse Committee of the Royal Astronomical Societies, of which I am Chairman, has received a grant of 1,000 pounds for the observation of a total eclipse of the Sun in May of next year, on account of its exceptional importance. Under present conditions, the eclipse will be observed by very few people. Prof. Eddington is peculiarly qualified to make these observations, and I hope the Tribunal will give him permission to undertake this task."

The letter did the trick. The local tribunal said it was convinced that Eddington was a true conscientious objector *and* that his work was of vast importance "not only to this country but to the world—to knowledge generally." The tribunal granted a twelve-month exemption to Eddington, provided that he continue his studies, particularly of the forthcoming eclipse. During the darkest days of World War I, with the German army shelling Paris, Eddington and his team of British astronomers got the official go-ahead to test a strange new theory proposed by a German-born scientist who published his work behind enemy lines. Not only that, but if Einstein was right, it would topple Newton, a founding father of modern scientific thought and a national hero in Britain.

The astronomers would journey thousands of miles by steamer to remote parts of the Amazon and Africa to observe the brief dimming of the Sun. Eddington and the Northamptonshire clockmaker Edwin Cottingham were bound for Príncipe, a Portuguese-owned island off the west coast of Africa, while Andrew Crommelin and Charles Davidson of the Royal Greenwich Observatory would travel to Sobral, in northern Brazil.

The trip, Eddington said, was the astronomical adventure of his lifetime. It was also an extraordinarily personal adventure for him, engaging his core beliefs as a Quaker. Eddington saw the eclipse expedition as a venue in which humanitarianism could

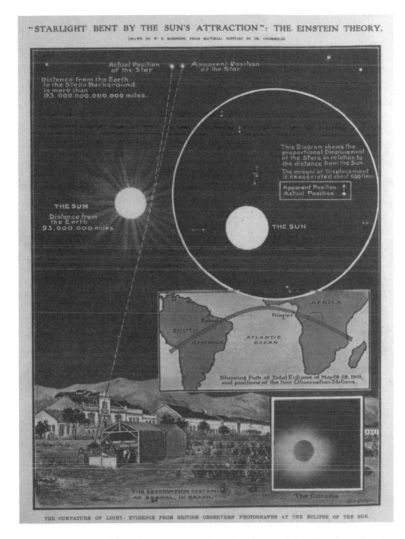

This drawing from the November 22, 1919, edition of the *Illustrated London News* shows several features of Arthur Eddington's 1919 solar eclipse expedition, the results of which confirmed a key prediction of Einstein's general theory of relativity and made Einstein a celebrity overnight. *(Illustrated London News.)*

claim victory over war and hatred. In leading the 1919 expedition, he shared the motivation of the Quakers who, after World War I, organized refugee camps and braved the blockade by the Allied powers to feed starving German children. Those Quakers, derided in England as "Hun lovers," became known as adventurers. Eddington, in his effort to measure the bending of light, saw himself in a similar light.

Before the journey could begin, Eddington and his colleagues had to choose and prepare the instruments they would take to Príncipe and Sobral. It was impossible to get work done before the armistice. In November 1918 the only workman available at the Royal Observatory Greenwich was the mechanic; the carpenter had not yet been released from military service. Scheduled to set sail in March 1919, the team had just a few months to get ready.

There was much to do. In January and February, Eddington made nighttime observations of stars in the Hyades cluster. He did so because he needed to note the positions of those stars at a time when the Sun was not present. That way, he'd be able to compare the positions of those same stars when the Sun *was* present, during the eclipse. The amount by which the stars appeared to shift their location due to the presence of the Sun would determine whether Einstein or Newton was right.

While Eddington observed the stars, Crommelin focused on gathering the tools. He did what he could to make sure the instruments—astrographic telescopes, designed to photograph large areas of the sky, along with a set of reflecting mirrors—were in excellent working order. They would have to be if the team was to have any hope of measuring the tiny deviations in the path of starlight predicted by Einstein. The stars would have a maximum deflection of just 4.9 ten-thousandths of a degree—equivalent to the angle subtended by a quarter viewed from two miles away. That would require the astronomers to measure a shift in the path of starlight on their photographic plates no bigger than the period on a printed page.

The astronomers would not transport entire telescopes, with their complex and fragile moving parts. Instead, just the objective lenses of the astrographic instruments were shipped. The lenses were placed at one end of a hollow steel tube. During the eclipse, the other end would hold the photographic plates.

To maximize mechanical stability, the tubes would lie horizontal and motionless. It would be the job of a separate, rotating mirror called a coelostat to continuously reflect light from the eclipse into the tube as Earth rotated, keeping the photographic plates centered on the blacked-out Sun and surrounding stars. No electric motors were needed: the mechanism controlling the rotation of the coelostat was driven by a falling weight.

Two 16-inch coelostats were shipped, one for each observing site. But their operation had proven troublesome during previous eclipse observations. Just before the expeditions were set to depart England, Father Aloysius Cortie, a Jesuit astronomer and director of the Stonyhurst College Observatory, provided the scientists with a backup—an eight-inch coelostat that would be used to reflect light into a four-inch telescope loaned from the Royal Irish Academy. Cortie had successfully used the small telescope to study the 1914 solar eclipse, and now Crommelin and Davidson would take it to Brazil.

On March 8, 1919, with Europe still technically at war (the Versailles peace treaty that would formally end the state of war between Germany and the Allied Powers would not be signed until June), the two teams set sail from Liverpool on the steamship *Anselm*.

The astronomers traveled together as far as the island of Madeira. After a farewell lunch on the island, Davidson and Crommelin reboarded the *Anselm* en route to Brazil. Eddington and Cottingham remained behind. With transatlantic steamer service only just having been resumed after the war and schedules still uncertain, the astronomers had to wait nearly a month for a Príncipe-bound ship to arrive.

Eddington, who thought nothing of going on seventy-mile bicycle rides, spent the time on Madeira mountain climbing; Cottingham could not keep up. As they toured the island, they viewed the ravages of the Great War. Parts of the town had been bombarded, and in Madeira harbor, the masts of two torpedoed ships stuck out of the water.

On the *Anselm*, Crommelin and Davidson reached the northern Brazilian state of Pará in March but decided to stay on the ship for another month since their accommodations at Sobral were not yet ready. They cruised a thousand miles of the Amazon, marveling at the luxuriant forests and the gorgeous plumage of the birds. The turbid yellow of the river contrasted with the color of two tributaries—the clear green of the Tapajós and the dark brown of the Rio Negro. In the town of Flores, Crommelin and Davidson hiked into the forest, the ground alive with troops of leaf-cutting ants, each carrying its green burden.

On Madeira, Eddington and Cottingham finally secured passage to Príncipe on April 9. The SS *Portugal* brought them to the island on April 23, five weeks before the eclipse. They found a tropical paradise of ancient rain forest, coffee and cocoa plantations, sandy beaches, and cloud-wreathed mountains. Principe also had a checkered history—plantation workers were descendants of slaves and forced paid laborers that the Portuguese had captured from the African mainland as recently as the 1870s.

To protect the equipment from the humid climate, laborers from a nearby cocoa plantation helped Eddington and Cottingham build waterproof huts. The scientists worked under mosquito netting and helped hunt monkeys that had been disturbing their equipment.

In mid-May, Eddington began taking practice images. The equipment worked well. But on the morning of May 29, after traveling thousands of miles to view the eclipse, Eddington and Cottingham awoke to a rainstorm. The storm ended two hours before totality but left behind intermittent clouds; observations of the stars the team needed to photograph would be touch and go. The eclipse would reach totality at 2:30 local time.

At about 1:30, Eddington and his colleagues began to get glimpses of the Sun, and at 1:55 they saw the crescent Sun through clouds, with large patches of clear sky appearing. The team raced through sixteen exposures, Cottingham giving the commands and attending to the driving mechanism of the coelostat and Eddington carefully changing the glass photographic plates to avoid shaking the telescope. Eddington never saw totality—he was too busy changing plates to look up. To Dyson, he telegraphed later that day: "Through cloud. Hopeful."

Eddington and Cottingham spent the next six nights developing the plates, two each night. During the days, Eddington analyzed the images. Although one plate showed a beautiful prominence—a hot filament of gas arcing from the solar disk—clouds covered the stars. Only two of the sixteen plates contained enough stars to measure the light bending,

Eddington had wanted to make all the measurements on the spot, but an imminent steamboat strike would have left him and Cottingham stranded on Príncipe for months. He and Cottingham took the first boat out, some two weeks after the eclipse, and arrived in Liverpool on July 14.

At Sobral, Davidson and Crommelin set up their eclipse station on a racecourse just in front of the house that served as their sleeping quarters. The course had a covered grandstand—and no horse racing to interfere with the observations. Bricklayers and carpenters were placed at their disposal to build supports for the telescopes and coelostats. Davidson and Crommelin also

A glass positive photograph of a total solar eclipse recorded in Sobral, Brazil, on May 29, 1919, during an expedition organized by astronomer Arthur Eddington to test Albert Einstein's general theory of relativity. *(Royal Astronomical Society.)*

had access to the first motorcar ever seen on Sobral, brought over from Rio expressly for their use.

Eclipse day began cloudy, but the sky cleared just before totality. As the Sun disappeared, leaving just the halo of the Sun's corona in view, a member of the team yelled, "Go!" An assistant immediately switched on a metronome, calling out every tenth beat to time the photographic exposures. Davidson and Crommelin took nineteen images with the main instrument, and eight with the backup telescope that Cortie had lent them. "Eclipse splendid," the team cabled to England later that day.

But the next day told a different story. After developing some of the images taken with the large astrographic lens, the astronomers were dismayed to find the plates out of focus. The Sun's heat

Eclipse equipment from Sobral, Brazil, one of two sites from which British astronomers observed the 1919 eclipse. The four-inch lens is in the square tube on the right, and the astrographic lens, which has a wide field of view, lies in the circular tube on the left. Mirrors that direct light into the tubes are driven by a mechanism that keeps the star image at the same position on the plates during an exposure. The mirror on the left was the suspected culprit in the poor quality of the images produced by the astrographic lens during the 1919 eclipse. *(Royal Astronomical Society.)*

had apparently caused the mirror to expand unequally. It would be difficult if not impossible to use the images to determine whether Einstein or Newton was right.

But the images taken with Cortie's telescope showed no such distortion. It was this instrument, the little telescope that could, that appeared to have saved the day.

The team left Sobral in June but returned in July to photograph the same group of Hyades stars during their first rising in the night sky, just before dawn, when the Sun was no longer present.

Davidson and Crommelin arrived back in England on August 25. Now both teams began the painstaking and sometimes tedious task of making measurements.

The researchers confirmed that the poor quality of the astrographic images taken on eclipse day at Sobral with the larger telescope were likely due to heating of the rotating mirror, as star images taken at night in July with the same lens did not show a similar distortion. The positions of stars in those out-of-focus images gave an average displacement of light of 0.93 arcseconds, close to the Newtonian light-bending value of 0.87. Einstein would have been wrong.

But during the analysis, it was decided to exclude all the flawed Sobral images, rather than just giving them lower statistical weight. Historians have accused Eddington of doing so in an attempt to force a solution that would prove Einstein right. But a review of correspondence among the researchers by Daniel Kennefick, a historian and astrophysicist at the University of Arkansas, revealed it was almost certainly Dyson, known to be much more of an Einsteinian skeptic, who made that decision.

Seven of the eight plates taken with Cortie's backup telescope at Sobral were of good quality. They contained seven stars whose positions could be analyzed. Significantly, those images showed that the deflection of starlight was largest for stars closest to the edge of the Sun and smallest for those farthest away, an effect that only Einstein's theory predicted.

In the end, Eddington's team reported two numbers. From the Sobral images, the researchers found a value for light bending of 1.98 arcseconds plus or minus 0.12 arcseconds. The Príncipe observations, based on fewer images, yielded a value of 1.61 arcseconds with a larger uncertainty, plus or minus 0.3 arcseconds. The findings, though based on limited data, agreed with Einstein's prediction.

Einstein learned about the preliminary results in the summer. On September 27, he wrote to his mother, who was dying of cancer: "Today some happy news. H. A. Lorentz telegraphed me that the

English expeditions have locally verified the deflection of light by the Sun."

The team announced their findings in London on November 6, at a joint meeting of the Royal Society and the Royal Astronomical Society that had been convened expressly to learn of the eclipse results. Most of the astronomers in the audience did not know what was to be announced. J. J. Thomson, president of the Royal Society and a Nobel laureate in physics for his discovery of the electron, presided over the meeting, held in the Great Hall of the colonnaded Burlington House.

"The whole atmosphere of tense interest was exactly like that of the Greek drama," wrote mathematician and philosopher Alfred Whitehead, who was present in the packed room. "There was a dramatic quality in the very staging:—the traditional ceremonial, and in the background the picture of Isaac Newton to remind us that the greatest of scientific generalizations was now, after more than two centuries, to receive its first modification. Nor was the personal interest wanting: a great adventure in thought had at length come safe to shore."

Dyson spoke first, outlining the history of the expeditions. In his summary of the results, he focused on the Sobral findings—the problem with the sixteen-inch coelostat and the high quality of the photographic plates made with the smaller telescope. He concluded: "After a careful study of the plates I am prepared to say there can be no doubt that they confirm Einstein's prediction. A very definite result has been obtained that light is deflected in accordance with Einstein's law of gravitation."

Crommelin was up next, filling in details about the Sobral observations, including the comparison of the positions of stars recorded during the eclipse and their position two months later from the Sobral site, when the Sun was not in the same part of the sky.

Then Eddington took the floor. He described the quality of the two images taken at Príncipe that contained enough stars to be analyzed. Combining the results of the two expeditions pointed to Einstein's value for the deflection, 1.87, rather than the Newtonian prediction of half that value, Eddington said. "For the half effect we have to assume that gravity obeys Newton's law; for the full effect which has been obtained we must assume that gravity obeys the law proposed by Einstein. This is one of the most crucial tests between Newton's law and the proposed new law."

Eddington added a caveat: "This effect may be taken as proving Einstein's *law* rather than his *theory*." Eddington meant that although he believed the observations confirmed the light bending predicted by Einstein, the study did not prove Einstein's claimed source for the bending—the curvature of space-time.

The measurements had large uncertainties, and some scientists in the audience were skeptical. The physicist Ludwik Silberstein proclaimed that although he was convinced that the observations proved the deflection of starlight, they did not convincingly demonstrate that the culprit had to be the curvature of space-time. "We owe it to that great man," he said, pointing to Newton's portrait, "to proceed very carefully in modifying or retouching his Law of Gravitation."

But Thomson spoke for those in the assembly who believed Einstein was right. "This is the most important result obtained in connection with the theory of gravitation since Newton's day and it is fitting that it should be announced at a meeting of the Society so closely connected with him," he said.

The next morning, November 7, the front page of the *Times* of London was full of stories about war and remembrance. It was only a few days before the first anniversary of the armistice, and King George V had just issued an invitation for all workers to take two minutes of silence out of their day to remember and honor "the glorious dead." But to the right of these stories appeared an

LIGHTS ALL ASKEW
IN THE HEAVENS

Men of Science More or Less Agog Over Results of Eclipse Observations.

EINSTEIN THEORY TRIUMPHS

Stars Not Where They Seemed or Were Calculated to be, but Nobody Need Worry.

New York Times headline from November 10, 1919, four days after Eddington presented his solar eclipse results to the Royal Society. *(New York Times.)*

article about rebirth and renaissance. In a triple-decker headline, the normally staid *Times* wrote: "Revolution in Science / New Theory of the Universe / Newtonian Ideas Overthrown."

The news set off a chain reaction around the globe. The *New York Times* followed suit with a front-page story on November 10: "Lights All Askew in the Heavens . . . Einstein Theory Triumphs."

In Berlin, the originator of the new theory, the forty-year-old Einstein, awoke as usual in the apartment he shared with his second wife and two stepdaughters. Berlin was still consumed with the privations of war and its aftermath, including scarcities of food and fuel for heat, but overnight Einstein had become the first science superstar.

He wrote to a colleague that he felt sure his sudden fame would soon die down. He was wrong. Einstein's celebrity would endure not just for days or weeks but throughout his lifetime and beyond, just as his theory of gravitation, a century later, continues to open new and unexpected windows on the birth and life of the cosmos.

....

DEEPER DIVE: A History of Light Bending

Einstein was not the first to suggest that light is bent by gravity. Newton himself raised the possibility. At the end of his 1704 treatise, *Opticks,* the sixty-one-year-old scientist asked a series of questions that he felt he did not have time to investigate but hoped others might. Query number one read: "Do not Bodies act upon Light at a distance, and by their action bend its Rays; and is not this action . . . strongest at the least distance?" Newton was not invoking curved space-time when he asked the question. He theorized that light was made of tiny particles (or corpuscles, as he called them) that had mass and therefore would respond to gravity.

In 1783, the English astronomer and clergyman John Michell took Newton's notion of light bending to an extreme and calculated that some objects have such strong gravity that no light could escape—a black hole, in modern parlance. A few years later, in 1796, French mathematician Pierre-Simon Laplace had a similar idea and calculated the mass of such an object: "The gravitation attraction of a star with a diameter 250 times that of the Sun and comparable in density to the earth would be so great no light could escape from its surface. It is therefore possible that the largest luminous bodies in the universe may, through this cause, be invisible."

But it wasn't until the beginning of the nineteenth century that German astronomer Johann Georg von Soldner published a calculation showing the amount of bending experienced by starlight grazing the Sun. Von Soldner calculated that the bending would make it seem as if the star had shifted its position by 0.875 arcseconds—the tiny angle subtended by a quarter when viewed from a distance of 5 kilometers. The English physicist Henry Cavendish appears to have made a similar calculation earlier, around 1784, but his results were never published.

Newton's particles of light—what physicists now call photons—are in fact massless, so the calculations by Soldner and other scientists did not stand the test of time. But when Einstein did his first calculation of starlight deflected by the curvature of space-time around the Sun, he obtained the same numerical value as Soldner. (He was unaware of Soldner's result, which in any event was not based on the same physical concepts.) In 1915, Einstein realized he had not included the curvature of space, only time, in his earlier work, and the deflection was actually twice what he had calculated. Luckily, he corrected the error before successful observations were conducted.

But the story does not quite end there. In the early 1920s, German experimental physicist and Nobel Prize winner Philipp Lenard, an anti-Semite and later Nazi sympathizer who was jealous of Einstein's very public success, accused Einstein of plagiarizing the results of Soldner, an Aryan scientist. In 1921, Lenard republished Soldner's paper with a long introduction in which he argued that Einstein's work was not original and that Soldner's calculation was in fact correct. His diatribe was part of the effort to denigrate what the Nazis would call "Jewish physics." Lenard was joined in his efforts by Johannes Stark, another Nobel Prize winner in physics.

A year earlier, in August 1920, a large anti-relativity rally had been held in the auditorium of the Berlin Philharmonic. A month later, Einstein and Lenard had gone head-to-head in an acrimonious and highly publicized debate on relativity in Bad Nauheim, Germany.

Like today's accusations of fake news, the fake charges against Einstein went viral. They spilled over to the United States when Arvid Reuterdahl, the dean of the engineering school at the College of St. Thomas in Minnesota, repeated Lenard's claims. His criticisms were published in detail in the *Minneapolis Tribune*. Ultimately, the furor died down when experiment after experiment confirmed the light bending predicted by Einstein.

DEEPER DIVE: A Modern-Day Solar Eclipse

The parking lot was packed with telescopes instead of cars. It was nearing 11:40 A.M. local time in Casper, Wyoming, on August 21, 2017, and hundreds of people had gathered to witness a total solar eclipse sweeping across the continental United States. Among them was astronomer Bradley Schaefer, who was using a small, computer-guided telescope to reproduce Arthur Eddington's famous solar eclipse experiment of 1919 that confirmed Einstein's general theory of relativity.

As the Moon passed over the Sun, shadows cast by trees took on strange shapes; the gaps between leaves acted like pinhole cameras, projecting crescent-shaped images of the eclipsed Sun onto the ground. The purple sagebrush and black-eyed Susans framing the parking lot disappeared from view. As the sky darkened, the temperature dropped and the world hushed: birds stopped singing, preparing for nightfall.

At totality, a red-hued prominence—a loop of hot gas—protruded from one edge of the Sun's darkened disk. "Oh, boy!" exclaimed Schaefer. You'd never have guessed that he had witnessed a score of other solar eclipses.

In the end, atmospheric distortion and other factors prevented Schaefer from clearly measuring the light-bending, although some other observers succeeded. But for this writer, who had never seen a total solar eclipse, the two-and-a-half-minute event provided indelible proof that nature, not man or woman, rules.

"Oh, boy!" indeed.

4

EXPANDING THE UNIVERSE

EVEN BEFORE Einstein arrived at the final equation for his gravity theory, he had applied general relativity to several phenomena in the solar system—the unaccounted-for precession of Mercury's orbit, the deflection of starlight passing near the Sun, and a predicted shift, due to gravity, in the color of sunlight.

But early in 1917, a little more than a year after his presentation to the Prussian Academy of Sciences, Einstein decided to tackle the entire universe. Finding out if his theory could accurately describe the cosmos at large "was a burning question," he told a friend and colleague, the Dutch astronomer Willem de Sitter. He needed to know, he said, whether his theory would succeed or fail.

Cosmology before Einstein had largely been the domain of philosophers and theologians. When it came to the universe, scientists simply did not have the mathematical tools (and large enough telescopes) to go beyond mere speculation.

Now Einstein was providing an entire mathematical arsenal—not just for himself but for a group of researchers eager to explore the nature of the universe. His theory would utterly transform how humans viewed the cosmos; his equations, combined with observations, would finally unveil the universe's vital statistics—its age, shape, composition, and mass.

But when Einstein began his study in 1917, he came face-to-face with a big problem. His equations told him that the universe could not stand still; it was either expanding or contracting. Yet observations indicated the cosmos was doing neither. Stars appeared to be nearly stationary, making it look as though the universe were a giant couch potato, remaining fixed and immutable for eons.

For once, Einstein bowed to the observational data gathered by astronomers. He hated to tinker with his perfect, mathematically elegant theory. But to save the universe in his theory from collapsing or expanding, he inserted a fudge factor: a constant that was denoted by the Greek letter lambda (λ) and would come to be called the cosmological constant. The cosmological constant acted as a kind of cosmic push, just strong enough to balance gravity's usual ability to pull material together. Although the constant could be viewed as endowing empty space with an unusual kind of energy, Einstein treated the term as a purely mathematical construct, something he inserted merely to keep the universe in check. (The constant was small enough not to interfere with the theory's successful predictions about the solar system.)

As he told his friend Felix Klein: "The new version of the theory means, formally, a complication of the foundations and will probably be looked upon by almost all our colleagues as an interesting, though mischievous and superfluous stunt, particularly since it is unlikely that empirical support will be obtainable in the foreseeable future."

The equation now read:

$$R_{\mu\nu} - g_{\mu\nu}R/2 - \lambda g_{\mu\nu} = -kT_{\mu\nu}$$

However, others who began exploring his equations, including de Sitter, weren't so sure Einstein had kept the universe still. De Sitter incorporated Einstein's cosmological constant and for simplicity began with a universe that was devoid of matter. That model, published in 1917, did indeed lead to a stationary cosmos. But when the English astronomer Arthur Eddington sprinkled matter into de Sitter's model, the particles flew apart—the universe was expanding.

In 1922, the Russian cosmologist and mathematician Alexander Friedmann found an even greater contradiction. Friedmann had come to relativity theory a bit late. During World War I, when scientific communication with other countries was impossible, he had volunteered with the Russian air force, applying his mathematical abilities to direct bombing missions. The chaos of the Russian Revolution of 1917 also prevented Einstein's theory from reaching Russia in a timely fashion; shipment of foreign scientific journals didn't resume until 1921.

Friedmann began with two assumptions—that the universe had a uniform distribution of matter and that it looked the same in all directions. He found that, depending on the numerical value of Einstein's cosmological constant, a static universe was just one of several possible scenarios allowed by Einstein's theory. The universe could also expand, contract, or oscillate between contraction and expansion. In describing the last scenario, he wrote, "The universe contracts into a point (into nothing) and then increases its radius from the point up to a certain value, then again diminishes its radius of curvature, transforms itself into a point."

The time it took for his oscillating universe to transition from contraction to expansion, it turned out, would be surprisingly close to the age of the universe derived years later from the Big Bang model. Although Friedmann's work provided the framework for a dynamic, expanding universe within the general theory of relativity, he did not live to see it: Friedmann died of typhoid fever in 1925 at age thirty-seven.

Einstein, for his part, remained unmoved. In reviewing Friedmann's 1922 paper, he initially called it suspicious and incorrect. He soon reversed his opinion, acknowledging that the article was in fact mathematically sound, but still believing it was physically irrelevant.

But new observations were making it harder to refute that the cosmos was growing. Using larger telescopes that peered deeper into the universe than ever before, along with improved spectrographs, instruments that separate starlight into its component colors, astronomers discovered that the universe was vastly bigger than anyone had imagined.

At Lowell Observatory in Flagstaff, Arizona, the young astronomer Vesto Slipher used a spectrograph to measure the shift in wavelength of light from the Andromeda nebula, which had been assumed to be a group of stars and dust in the Milky Way. Light appears shifted to the bluer part of the spectrum if a body is moving toward Earth and shifted to the redder part of the spectrum if it's receding. In 1912, Slipher measured that Andromeda had such a large blueshift—indicating a high velocity toward Earth—that the nebula must lie outside the Milky Way. Andromeda was another entire galaxy! By the mid-1920s, Slipher had measured the shifts of some forty-one different galaxies, most of them receding from Earth at high speed.

Meanwhile, at the Carnegie Institution's Mount Wilson Observatory near Pasadena, California, astronomer Edwin Hubble was having a similar epiphany. Seated in the observing cage of the 100-inch Hooker telescope, the world's largest telescope from 1917 to 1949, Hubble measured the distances to many of the nebulae and found that they were much farther away than the estimated size of the Milky Way. He too was forced to conclude that they had to lie outside our galaxy and were indeed galaxies in their own right.

Hubble got his first clue in 1923, while observing the Andromeda nebula—the same nebula for which Slipher had measured a blueshift. Within the nebula, he identified a Cepheid variable, a star that regularly waxes and wanes in brightness. Hubble and his colleagues were elated because they knew the Cepheid would enable them to accurately measure the distance to Andromeda.

For that knowledge, Hubble owed a debt to the American astronomer Henrietta Swan Leavitt of the Harvard College Observatory, who more than a decade earlier had found a special connection between the brightness of a Cepheid and its period of pulsation: the brighter the Cepheid, the longer its period. That relationship meant that if astronomers measured the duration of the pulse period, they could determine the true brightness of the star. Just as a lightbulb appears fainter when held at a great distance, a star that lies farther away looks dimmer than it really is. Comparing a Cepheid's true brightness with its dimmer appearance in the sky, astronomers could then calculate how far away the star must be. And in the case of the Cepheid in Andromeda, the distance confirmed that the nebula was indeed another galaxy.

In 1927, the French priest and physicist Georges Lemaître independently found a solution for an expanding universe using Einstein's theory. Because he had access to telescope observations that Friedmann did not, Lemaître went further than his predecessor had. Lemaître asserted that a galaxy's light is stretched in frequency *by the expansion of space itself.* The longer the light's journey, the more the universe had expanded and the greater the light's redshift, he predicted. He backed up his claim by combining redshift measurements for spiral galaxies, compiled by Slipher, with the distances to those galaxies, measured by Hubble. The data also allowed Lemaître to derive an early estimate for the rate at which the universe was expanding.

In 1929, Hubble confirmed Lemaître's work, using observational data to formally demonstrate the linear relationship between speed (redshift) and distance. The speed at which galaxies were receding from Earth was proportional to their distance. Those that resided twice as far from Earth were speeding away twice as fast, those that were four times more distant were fleeing with four times the speed, and so on.

This would be true not only from Earth's vantage point but from any point in the universe. Think of a partially inflated balloon, imprinted with tiny dots. When someone adds more air to the balloon, each dot (a stand-in for a galaxy) moves away from every other dot. Moreover, it does so in proportion to its original distance from each dot. The farther apart the two dots were when the balloon was only partially inflated, the greater their separation as the balloon expands.

Even as evidence for an expanding universe mounted, Einstein refused to budge. When he and Lemaître met at a conference in 1927, Einstein told him the work was mathematically sound but that his grasp of physics was "abominable."

However, in 1930 Einstein began to have a change of heart. Eddington demonstrated that Einstein's cosmological constant was an unstable solution. The constant did keep the universe poised between expansion and contraction, but it was balanced on a knife's edge. If the constant was even infinitesimally larger than the preferred value, the universe would expand; if it was only slightly smaller, the universe would collapse. That meant the solution wasn't viable.

Two great scientists conversing at the University of Cambridge in 1930. During the solar eclipse of May 29, 1919, Arthur Eddington (*right*) confirmed that starlight passing near the Sun is bent by an amount predicted by Einstein's general theory of relativity. *(Royal Astronomical Society.)*

Early in 1931, Einstein visited Hubble at the Mount Wilson Observatory. He and Hubble rode up the dirt road to the observatory in a Pierce-Arrow touring car. At the top, Einstein was like a kid with a new toy. He climbed into the telescope's cage, fascinated by all the instruments and dials. After the visit, he told the press he was now convinced the universe was expanding. He immediately began work on a model of cosmic expansion and retired his cosmological constant. But the constant would not stay dead forever.

In 1931, Lemaître published a new paper in which he again examined Einstein's equations but this time put the movie of cosmic expansion in reverse. If the universe was expanding now, it must have been smaller in the past. Wind the clock back further, and the universe would have been infinitesimally small. Perhaps, he suggested, it had all begun with an explosive, primeval atom.

"The evolution of the world can be compared to a display of fireworks that has just ended: some few red wisps, ashes and smoke," he wrote. "Standing on a well-chilled cinder, we see the slow fading of the suns, and we try to recall the vanished brilliance of the origin of worlds." That was a poetic way of describing what British astronomer Fred Hoyle, who strongly disliked the expanding-universe model, derisively called the Big Bang. The name stuck.

In 1948, around the same time that Hoyle coined the term, Russian-born cosmologist George Gamow, who had briefly studied with Alexander Friedmann, described a universe that began its expansion from a hot, dense state. But unlike the primeval atom envisioned by Lemaître, Gamow's infant universe consisted mostly of radiation. In a short follow-up paper, two of Gamow's young colleagues, Ralph Alpher and Robert Herman, calculated that if the cosmos had indeed started off with a Big Bang that had sparked the expanding universe and forged several of the lightest elements, there ought to be some heat left over from the explosive event. Alpher and Herman predicted that a sensitive enough radio telescope should be able to detect that relic heat, which the researchers calculated to have a temperature of about 5 kelvins, or 5 degrees above absolute zero.

By the 1960s, most researchers had forgotten their prediction about a background glow. Independently, Robert Dicke of Princeton University and his colleagues arrived at a similar conclusion and in the early 1960s began searching for the left-over heat, using a radio-wave detector mounted on the roof of a Princeton building.

Around the same time, Arno Penzias and Robert Wilson of Bell Laboratories in Holmdel, New Jersey, were cataloguing all known sources of radio emissions from space in an effort to improve satellite communications. They soon encountered something strange: some radio static persisted no matter where in the sky they pointed their horn-shaped radio antenna or what time of day they made their measurements. Pigeons had taken to roosting in the antenna, and Penzias and Wilson at first thought that heat from the bird droppings might be the source of the static. Booting the pigeons and scrubbing out the droppings didn't eliminate the noise, however.

Ultimately, the researchers had to conclude that a faint microwave hiss bathed the entire sky. They didn't realize they had discovered the cosmic microwave background (CMB)—the Big Bang's leftover heat, the ancient radiation that first streamed freely into space when the cosmos was about 380,000 years old.

"Boys, we've been scooped!" said Dicke when he heard the news. Penzias and Wilson won the Nobel Prize in Physics for their serendipitous discovery.

In the early 1970s, several theoretical physicists, including Jim Peebles of Princeton and Rashid Sunyaev and Yakov Zel'dovich in Russia, built upon that discovery of the CMB. Combining a model of the expanding universe derived from the general theory of relativity with the behavior of light in the hot, early universe, they realized that the CMB should not have an entirely smooth, uniform glow. A perfectly uniform and smooth CMB could not have produced the lumpy universe of today, with its vast networks of galaxy clusters interspersed with giant voids. So if microwave detectors examining the CMB were sensitive enough, they ought to discern the seeds of those structures—places where matter is slightly more compressed than average, indicated by a slightly higher CMB temperature, and places where the density is slightly less, indicated by a slightly lower temperature. The hot and cold spots *had* to be there.

But by the early 1980s, physicists had more sensitive detectors and they still hadn't seen them. Could general relativity be wrong? Was the Big Bang explanation not right?

At that point, Peebles made what might have seemed an outlandish suggestion. He proposed that most of the matter in the universe was invisible and interacted only through its gravity. Since it did not interact with light, this invisible material, or dark matter, would generate smaller lumps in the CMB than ordinary matter; that would explain why no one had yet seen evidence of it.

The hypothetical dark matter Peebles invoked had been posited here and there for decades, but most scientists thought the idea was ludicrous. That might have been because one of the first people to suggest it was a brilliant but abrasive astrophysicist named Fritz Zwicky. His personality didn't exactly endear him to his colleagues. He once called co-workers at the Mount Wilson Observatory "spherical bastards," because, he said, they were bastards any way you looked at them.

In 1933 Zwicky examined a nearby group of galaxies called the Coma cluster and found that individual galaxies in the cluster were moving so rapidly that the gravity exerted by the visible parts of the cluster was too weak to keep the Coma intact. How was that possible? Zwicky's solution: all the visible material in the cluster accounted for a tiny fraction of the cluster's total mass. The rest, which could not be seen, he called *dunkle Materie*— German for "dark matter."

By the 1970s, Zwicky's crazy idea didn't seem so crazy. Astronomer Vera Rubin of the Carnegie Institution in Washington, D.C., working with Kent Ford, had been measuring the speeds of stars in different parts of spiral galaxies. They knew that the highest concentration of visible mass resided at the cores of the galaxies. They therefore assumed that stars at the outskirts of the galaxies, far from the core, would orbit more slowly than stars closer in, just as the solar system's outer planets orbit the Sun more slowly than the inner planets. But Rubin found that the stars at the outer edge rotated just as rapidly. She concluded that the galaxies had to lie inside a halo of dark matter—and that there had to be ten times as much of it as there was visible material.

Peebles needed dark matter just as much as astronomers Rubin and Zwicky had. And in 1992, NASA's Cosmic Background Explorer satellite finally found evidence of the tiny hot and cold spots in the CMB that one would expect if dark matter ruled the universe. Since 1992, numerous experiments on the ground and in space, notably NASA's Wilkinson Microwave Anisotropy Probe and the European Space Agency's Planck satellite, have explored the fluctuations in detail. From the nature of the temperature variations, astronomers have gleaned the size and shape of the universe, properties that can be traced back to space-time curvature and predictions from general relativity.

Astronomers have also found evidence that the ingredients of the universe are both more mysterious and darker than most researchers had suspected—and that Einstein may have been too hasty when he abandoned the cosmological constant. That story harks back to the 1990s and ran along two seemingly disparate tracks.

Observations of the size of the hot and cold spots in the CMB had revealed that the universe was flat on a large scale— that is, the angles of a triangle would always add up to 180 degrees and parallel lines would never meet. Those observations dovetailed nicely with a leading model of the early universe called inflation. The theory posited that the young cosmos underwent a humongous growth spurt, enlarging from atomic size to the diameter of a soccer ball in a minuscule fraction of a second. Inflation, which offered an explanation for how giant galactic structures grew from the tiny lumps in the CMB, would also make the universe flat.

A flat universe would require that the total allotment of energy and matter in the cosmos equaled a certain critical density. But measurements of the actual amount of mass in the universe— both visible matter and dark matter—had come up drastically short. There simply wasn't enough matter of any type to keep the universe flat. Where was the missing stuff?

The answer would come from an entirely different set of studies. In 1998, Adam Riess, then a postdoctoral fellow at the University of California, Berkeley, was about to leave for his honeymoon when he emailed his colleagues that the universe appeared to be almost completely dark and repulsive. Fortunately, he was talking about a matter of gravity, not his view of married life. Riess was part of a team viewing distant supernovas to study the expansion of the universe. The researchers used supernovas like standard lightbulbs to measure distance, in a manner similar to the way Hubble had used the brightness of Cepheid stars. Because supernovas are much brighter than Cepheids, however, they can be used to study cosmic expansion at much greater distances.

Riess's team, led by Brian P. Schmidt, then at the Mount Stromlo and Siding Spring Observatories in Australia, had expected that cosmic expansion, driven by the heat released during the Big Bang, ought to have been decelerating ever since the moment of initial inflation, slowed by the mutual gravity of all the matter in the cosmos. But that's not what the astronomers, along with a rival group led by Saul Perlmutter of the Lawrence Berkeley National Laboratory in California, had found. Instead of slowing, cosmic expansion was speeding up. Gravity had somehow transformed from an attractor to a repeller, forcing matter to fly apart at an ever-faster rate. It was as surprising as throwing a ball up in the air and, instead of seeing it come back down, finding that it just keeps on traveling upward faster and faster.

Was Einstein's theory wrong, or was there something missing from astronomers' description of the universe?

After some soul-searching, astronomers and cosmologists had to accept the notion that gravity had a flip side. Some kind of invisible, mysterious energy fills the universe, turning gravity's pull into a cosmic push. It's derived straight from Einstein's theory. Cosmologists call this mysterious force dark energy. But it could just as easily be called the cosmological constant. It pervades all of space, doesn't dilute as space-time expands, and accounts for about 68 percent of the universe's mass and energy.

In doing so, dark energy balanced the cosmic ledger book. It provided the critical density of mass and energy that the CMB studies and inflation dictated must be present to keep the universe flat.

More than two decades after dark energy's discovery, scientists still have no killer theory to explain its existence. Cosmic acceleration has been variously proposed as arising from the energy of empty space, as a leftover from the epoch of inflation at the birth of the universe, or as a result of gravity leaking away into extra, hidden dimensions.

Dark energy, or repulsive gravity, was always present, but at the beginning of the cosmos it had little impact. The youthful universe, though expanding, was relatively compact and dense. The high density enabled gravity to reign supreme. But as the universe continued to expand, the density of matter decreased, along with gravity's tug. In contrast, dark energy is a feature of space that does not diminish as the cosmos expands. It is a constant throughout both space and time: about a hundred-millionth of an erg per cubic centimeter. Eventually, about 5 billion years ago, the cosmic push of dark energy won the tug-of-war against gravity's pull, and cosmic expansion began to accelerate.

If dark energy does have a constant density, spread evenly throughout space, then it would indeed resemble the cosmological constant, that feature that Albert Einstein inserted into his theory of gravitation in 1917. After Einstein conceded that the universe was indeed expanding, he disowned the constant, reportedly calling it his greatest blunder. But he may have been right after all.

5

BLACK HOLES
AND TESTING
GENERAL RELATIVITY

S OON AFTER Einstein completed his general theory of relativity, a handful of physicists began exploring how to apply
his work. Investigating the theory wasn't easy, however. The
mathematics was devilishly complex. The equation Einstein presented to the Prussian Academy of Sciences on November 25, 1915,
was elegant, but it symbolized ten coupled, nonlinear equations.
Each equation dealt with all four dimensions (three of space and
one of time). Einstein himself had only found approximate solu-
tions to his suite of equations.

Yet less than a month after Einstein's presentation, he received
a telegram from the German physicist Karl Schwarzschild, who
told Einstein he had found an exact solution. Schwarzschild
wasn't contacting him from a university; though he was over forty,
he had enlisted in the German army and was stationed at the Russian front. Despite the heavy gunfire, he told Einstein, he had

found time to take a respite from the war and "take this walk into this your land of ideas."

In a manuscript he mailed to Einstein, Schwarzschild had worked out a solution to Einstein's complicated equations for the space-time curvature outside a relatively simple object: a stationary, spherical mass, such as a non-rotating star. Einstein was so impressed by his result that he presented Schwarzschild's paper before the Prussian Academy, and a month later it was published.

But one aspect of the solution disturbed Einstein. Schwarzschild had found that if he squeezed the mass of his spherical system into a small enough radius, the system would undergo a catastrophic gravitational collapse from which even light could not escape. The material would crunch down to a point of infinite density and the equations of the general theory would go berserk. Schwarzschild had used relativity to describe what astronomers now call a black hole.

As surprising as Einstein found general relativity's implications for the universe on the largest scale, he was equally resistant to what his theory revealed about the universe on the smallest scale. He had trouble believing that conditions in nature could create a region where matter would be compressed to a vanishingly small volume, space-time would curve infinitely, and the laws of gravity would break down.

Yet Schwarzschild had even developed a formula for the radius (R) at which gravitational collapse would occur: $R = 2GM/c^2$ where G is the gravitational constant, c is the speed of light, and M is the mass of the object. For the Sun to become a black hole, all of its mass would have to be packed into a ball with a radius of about 3 kilometers. For Earth, the Schwarzschild radius is a little over one-third of an inch (8.7 millimeters), the size of a small marble.

Like Einstein, Schwarzschild viewed his results as a mathematical curiosity that had no physical reality. He too thought it unlikely that nature could exert the enormous force required to compress a star into a Schwarzschild radius.

But Schwarzschild didn't pursue his research further. While at the Russian front, he began suffering from pemphigus, a rare and painful skin disease that ravaged his immune system. He died on May 11, 1916, only months after his first telegram to Einstein.

Other scientists didn't show much interest in Schwarzschild's work in the years that followed. But the notion that gravitational collapse might actually occur in nature got a foothold in the 1930s. That's when astronomers began to wonder what happens to a star when it exhausts the fuel that keeps it burning and provides the outward pressure that opposes gravity's inward pull. With the fuel depleted, gravity would dominate, at least for a while, compressing the mass of the star.

Astronomers calculated that when our Sun runs out of fuel in 4 to 5 billion years, it will shed its outer layers and gravity will shrink the leftover core to the size of Earth. The compact core, known as a white dwarf, consists of a densely packed collection of atomic nuclei and electrons. Because electrons resist squeezing—quantum theory forbids any two electrons from occupying the same energy state—the compressed particles exert an outward pressure. For a star as massive as the Sun, that pressure is large enough to counterbalance gravity's pull. Gravitational collapse halts.

But what would be the fate for heavier stars? Would they collapse even further? That's what nineteen-year-old Subrahmanyan Chandrasekhar pondered as he sailed from India to England in 1930 to begin graduate work in astrophysics at the University of Cambridge. Aboard ship, he performed a calculation that would answer that question. Chandrasekhar found that for a white dwarf greater than about 1.4 times the mass of the Sun (corresponding to a star that began its life at least 8 times heavier than the Sun), electron pressure would be no match for gravity. The star would continue to collapse until its radius was no bigger than the size of a city, about 10 kilometers. Gravity would squeeze nuclei so tightly that electrons and protons would fuse to form neutrons. This compact body, a teaspoon of which would weigh a billion tons, is called a neutron star. Astronomers, including Arthur Eddington, balked at such a finding, but the work he did at age nineteen would garner Chandrasekhar a Nobel Prize in Physics.

In the late 1930s two researchers at the University of California, Berkeley, George Volkoff and J. Robert Oppenheimer, pushed gravity even further. They wanted to find out if neutron stars were truly the end stage of stellar collapse. Combining quantum theory with Einstein's theory of gravity, Oppenheimer and Volkoff calculated that if the initial mass of a star was sufficiently large, its neutron-star core would be too heavy to resist gravity and it would undergo a further, catastrophic collapse. (Recent studies indicate that any neutron star greater than 2.16 times the Sun's mass will succumb to gravity.) In a follow-up paper by Oppenheimer and his student Hartland Snyder, they described what would happen: "The star thus tends to close itself off from any communication with a distant observer; only its gravitational field persists." A black hole is born.

Imagine a voyager attached to the surface of the collapsing star, moving with it as the star continues to crunch down. Suppose that voyager sends light signals to an observer, far from the collapsing star. Due to the enormous space-time curvature, the light signals will shift to redder and redder wavelengths and take a longer and longer time to reach the faraway observer. In fact, it will appear to take an infinite amount of time for the star to reach the Schwarzschild radius. The distant observer can never know what happens after the shrinking star approaches the Schwarzschild radius because it will appear frozen in time. What's inside the Schwarzschild radius remains forever unknown to the outside observer—although theorists are trying several approaches to find out (see Chapter 6).

Oppenheimer and Snyder published their paper on September 1, 1939, the same day that World War II began. Oppenheimer would soon be leading the U.S. effort to develop the atomic bomb and never again wrote about black holes. The objects remained a mathematical curiosity until the early 1960s, when astronomers discovered quasars, compact objects whose blazing light was believed to be fueled by monster black holes. Today, black holes are thought to litter the universe. Observations suggest that a giant black hole lurks at the core of every large galaxy, where It governs the galaxy's formation and growth.

Black holes are not the only catastrophic collapse that general relativity must contend with. There's the Big Bang itself. If we imagine the expansion of the universe running in reverse, it suggests the cosmos began as an entity that was infinitesimally small—the universe confined to a point with infinite density. Such tiny spaces are, once again, the realm of the quantum.

Back in 1919, after observing the solar eclipse from Príncipe, British astronomer Arthur Eddington freely admitted that additional observations were needed to more accurately confirm Einstein's predictions of light bending. So astronomers traveled to southern Australia in 1922 for the next major solar eclipse opportunity, observing the event from a sheep farm. This time the skies were clear, and images revealed the apparent displacement of many more stars due to light bending.

Although the 1922 data confirmed Eddington's findings, they had about the same accuracy. The problem is that, observed from the ground, starlight gets blurred by Earth's turbulent atmosphere. So for decades not much progress was made in measuring the deflection of starlight. But the endeavor got a big boost with the 1989 launch of the European Space Agency's Hipparcos satellite, which flies high above Earth's light-obscuring atmosphere. Hipparcos pinned down the positions of stars to a thousandth of an arcsecond; with such precise measurements, the satellite was able to discern starlight being bent by both the Sun and the planets. The Gaia satellite, launched in 2013 by the European Space Agency, is poised to record the positions of billions of stars to an even higher degree of accuracy—about 20 millionths of an arcsecond—and is expected to see the effect of light bending by the Sun in every single one of its measurements.

But the most spectacular aspect of light bending emerged from a calculation Einstein did in 1912, three years before he completed the general theory of relativity. He showed that the most powerful magnifying lenses aren't on Earth but in the sky. Einstein described the properties of a gravitational lens, in which a massive foreground object (a cluster of galaxies, for example) bends light from a more distant body (a faraway star or galaxy) in such a way that the observer sees distorted but magnified images of that distant source: two or more distinct images, replete with rings, arcs, and brightness variations are all possible.

Einstein may have concluded that the effect could not actually be observed, notes science historian Tilman Sauer. Although Einstein and Eddington both remained cautious about the practical applications of lensing, in the late 1930s maverick astronomer Fritz Zwicky (see Chapter 4) calculated that it should be relatively easy to record clusters of galaxies acting as lenses for more distant bodies in the cosmos. Astronomers observed the first gravitational lens, a double image of a distant quasar, in 1979. Six years later, another team found four images of a different quasar, arranged in a cloverleaf pattern. In the years since, astronomers have found about 1,000 examples of gravitational lenses that produce multiple images of a celestial body.

Telescopes are time machines—when they look at faraway galaxies, they are seeing those galaxies as they appeared far

In this portrait by the Hubble Space Telescope of a massive cluster of galaxies known as Abel S1077, the mass of the cluster warps the surrounding space-time, acting as a lens that bends and brightens light from more distant galaxies that happen to line up with the cluster. Indeed, the stretched stripes that look like scratches on a lens are distant galaxies whose light is heavily distorted by the gravitational field of the cluster. *(N. Rose/ESA/Hubble & NASA.)*

back in time, since the light now reaching Earth left those bodies several billions of years ago. Today, the Hubble Space Telescope and other far-seeing telescopes use lensing to search for the most distant galaxies in the cosmos, which are some of the first galaxies ever born. Many of these baby galaxies would be invisible were it not for the magnified images generated by lensing.

Astronomers also use gravitational lenses to learn about the nature of the lens itself—in particular, how much dark matter it may contain. Dark matter, the mysterious material now believed to outnumber the amount of visible matter in the cosmos by nine to one, can't be seen. But this ghostly material betrays its presence through its gravity—how much it bends light from a distant body.

When observers spot a lens, they estimate the visible matter it contains and compare that number to the amount by which light is bent. The greater the light bending, the greater the mass that the lens must contain. Typically, the bending is much greater than the visible matter could generate—the rest must be due to dark matter. In this way, Einstein's theory is allowing astronomers to build a census of the invisible stuff throughout the universe.

Even the tiny lumps present in the cosmic microwave background, the heat left over from the Big Bang, get lensed. The lensing has aided the European Space Agency's Planck satellite in gleaning the information needed to determine such properties as the age, shape, and matter content of the universe.

Lensing is also shedding light on dark energy, the mystery force that pushes against gravity's pull and is accelerating the rate at which the universe is expanding. Dark energy and dark matter, the primary sources of gravity, are essentially in a tug-of-war: dark matter pulls material together, while dark energy tries to pry it apart. The amount of clumping in the universe is the direct result of that epic battle.

Over a five-year period that ended in 2018, astronomers leading a project called the Dark Energy Survey examined one-eighth of the sky, observing some 300 million galaxies that lie billions of light-years from Earth. By comparing subtle distortions in those images—the handiwork of lensing, primarily due to dark matter—the team can map out the distribution of dark matter when the universe was about half its current age. Comparing that clumpiness to how dark matter congregates at later times reveals whether dark energy's push has remained constant or changed.

In the decades during which visible-light tests of Einstein's theory languished, radio astronomy came to the forefront. By combining the acuity of several radio telescopes, astronomers could determine with great precision the positions of quasars—celestial objects that emit powerful, highly collimated beams of radio waves. Astronomers compared the positions of the radio sources when the Sun was and was not present in that part of the sky and measured deflections that were in agreement with Einstein's prediction.

Radio astronomy also contributed the so-called fourth test of general relativity (after the three tests suggested by Einstein himself: the precession of Mercury's orbit, light bending, and the gravitational redshift of light emitted by the Sun). In 1964, astrophysicist Irwin Shapiro proposed that general relativity could be tested by bouncing radar signals off another planet and measuring how long it took for them to arrive back at Earth. Shapiro argued that the round-trip journey would take longer when the planet was near the Sun compared to when the Sun was in a different part of the sky. He and his colleagues recorded radar reflections from both Mercury and Venus and found agreement with general relativity to the 5 percent level. In 2003, Italian astrophysicists measured the amount by which the Sun delayed radio waves sent from Earth to the Cassini spacecraft as it sped toward Saturn. That delay agreed even more strongly with relativity, to 20 parts in a million.

And still astronomers pursue even more stringent and sophisticated tests of Einstein's theory, including atomic-scale versions of the Leaning Tower of Pisa experiment (see "Deeper Dive: New Tests of Einstein's Theory"). With relativity passing every test with flying colors so far, why continue? Because sooner or later, the theory has to fail—and where it does, new physics may emerge.

....

DEEPER DIVE: New Tests of Einstein's Theory

For years after Einstein presented his theory of general relativity, experimental testing of his work lacked high precision. As Einstein's fame grew, relativity research languished. Many scientists viewed the theory as a backwater, intriguing but difficult to test and irrelevant to cutting-edge studies in physics and astronomy. But the technology that brought black holes and other oddities of relativity into mainstream physics also opened up a new era of testing.

In the summer of 1971, Apollo 15 astronaut David Scott performed a version of Galileo's apocryphal tower experiment in a place where air resistance could not confound the results—the Moon. Standing on the lunar surface, Scott held a hammer in his gloved right hand and a falcon feather in his left. With the world watching via live video, he dropped them from shoulder height. The feather did not flutter but plummeted to the surface just as fast as the hammer, striking at the same instant.

Mirrors left behind on the Moon by other Apollo astronauts and a Russian robotic mission in the late 1960s and early 1970s soon led to a more expansive version of the Galileo tower experiment. This time, the two masses were the Moon and Earth themselves. Although the Moon orbits Earth, both bodies are in free fall about the Sun. Not only do the two bodies have different weights—Earth is eighty-one times heavier than the Moon—but Earth is considerably denser, with an iron core that the Moon lacks. If the Sun's gravity accelerated one of the masses at a different rate than the other, it would show up as a tiny change in the distance between the Moon and Earth.

To measure the distance, astronomers relied on the lunar mirrors, known as retro reflectors. Light from Earth that strikes the mirrors is reflected back along exactly the same path that the beams took to reach the mirrors. By recording the round-trip travel time between Earth-based observatories and the mirrors, astronomers have pinned down the Earth-Moon distance—some 385,000 kilometers, on average—to within a few centimeters.

There's a particular advantage in testing relativity with heavy objects. Einstein's equivalence principle, in its strongest form, asserts that *all* forms of mass-energy experience the same acceleration in a uniform gravitational field. Some of this energy is in the form of gravitational binding energy, the glue that keeps the bits and pieces of an object together instead of flying apart. Gravitational binding energy should undergo the same acceleration that mass does, according to Einstein. But this is impossible to test in a ground-based laboratory because the test masses have so little binding energy. However, the heft of the Earth and the Moon make such measurements possible. Even so, Earth's gravitational binding energy is less than a billionth of its mass; the Moon's gravitational binding energy is an even smaller component of its mass. But it's enough to test the equivalence of these two forms of mass-energy.

Researchers from the University of Bologna analyzed nearly fifty years of lunar laser ranging measurements, looking for a displacement of the lunar orbit toward or away from the Sun that would arise if a fundamental principle of general relativity were violated: that all masses fall at the same rate in the Sun's gravity. The team reported that relativity's predictions agreed with the lunar data to a precision ranging from one part in 10 million to one part in 1 trillion. That improves upon ground-based measurements made at the University of Washington, where Eric Adelberger and his colleagues have used a highly sensitive, modern-day version of Eötvös's torsion balance to test the principle.

Even as each new test vindicates relativity, scientists press on, subjecting the equivalence principle to ever-stricter scrutiny. A failure of relativity could resolve the impasse that currently exists between quantum theory and gravity. The two appear to be utterly incompatible—quantum theory describes the material universe in terms of probabilities and uncertainties, while relativity assumes that space and time can be well defined down to the tiniest levels.

The clash between relativity and quantum theory hints that something is deeply wrong at the heart of physics. That's exciting because it suggests there may be something brand-new, waiting to be discovered. A breakdown of relativity could even reveal that a previously unknown force is at play in the universe.

To further test Einstein's theory, scientists have shipped their experiments to space, away from the seismic noise and other confounding effects on Earth. In 2016, the French space agency launched MicroSCOPE (Drag-Compensated Micro-Satellite for the Observation of the Equivalence Principle), a satellite that carried two concentric cylindrical shells, one made of a titanium alloy, the other platinum. As the satellite circled Earth in free fall, each shell should orbit in exactly the same way if the equivalence principle holds. If not, one of the shells would slip slightly ahead of the other, indicating that it was falling (accelerating) slightly faster. The movement of the shells was monitored, and to keep them centered within the satellite's housing, electrostatic forces were separately applied to nudge each cylinder as needed. Differences in the applied force required for each shell would indicate a violation of the equivalence principle. After analyzing data from 120 orbits about Earth, researchers in 2017 found no such violation, to a precision ten times better than what is possible with ground-based experiments. Information gleaned from the many thousands of orbits the craft took around Earth before the mission ended in the fall of 2018 promises to substantially increase that precision. To put the experiment in perspective, 1,900 orbits corresponds to a free-fall distance equal to half the separation between the Earth and the Sun.

One of the more far-flung tests of Einstein's equivalence principle involved a trio of stars 4,200 light-years from Earth. The triple system consists of three stars at the end of their lives—two white dwarfs (the collapsed remnants of stars similar to the Sun) and a neutron star (forged when a massive star explosively sheds its outer layers). Astronomers can track the motion of the system because the neutron star is a pulsar—it broadcasts radio waves like a lighthouse beam as it spins 366 times a second.

One of the white dwarfs and the pulsar form a close pair that together are in free fall around the second, outer white dwarf. If the inner white dwarf and the neutron star were to fall with different accelerations, it would show up as a change in the rate at which the pulsar's radio beams arrive at Earth. But measurements of the pair's motion reveal that despite differences in their mass and composition, the neutron star and the white dwarf fall at the same rate, to within 0.16 thousandths of a percent of each other. The finding confirms the equivalence principle in the extremely strong gravitational environment of a neutron star, where the full theory of general relativity is required. Because the stars are massive, the gravitational binding energy of each makes a sizable contribution to their total mass-energy. The experiment therefore provides a stronger test than the Earth-Moon system that gravitational binding energy and mass accelerate in the same way.

Astronomers have examined another prediction of relativity, gravitational redshift, in an even more extreme environment—the vicinity of the giant black hole, equivalent to the mass of four million Suns, packed into the core of the Milky Way. Since the early 1990s, two teams have monitored the motion of a star called S2, which orbits the galactic center in an elliptical path. One team has now caught the giant black hole stretching the light emitted by the star—the first time gravitational redshift has been observed near a black hole.

Researchers from the Max Planck Institute for Extraterrestrial Physics in Garching, Germany, measured the motion of S2 as it passed near the black hole in the spring of 2018. One instrument measured the component of S2's motion across the sky, while another recorded the component of motion toward and away from Earth. From these measurements the team deduced the amount by which the black hole's curvature of surrounding space-time shifts the light emitted by the star to redder wavelengths. The gravitational redshift agreed well with Einstein's theory. A rival team, led by

Illustration of Gravity Probe B, a satellite that measured the amount by which the rotating Earth drags space-time along with it. Einstein's general theory of relativity predicts that any massive, spinning body will generate this effect, known as frame dragging. By 2011, when researchers announced the final results from Gravity Probe B, a satellite that took forty years to get off the drawing board, other, cheaper missions had already beaten it to the punch. *(Courtesy Gravity Probe B Image and Media Archive, Stanford University.)*

Andrea Ghez of the University of California, Los Angeles, made similar measurements.

In 2015, a NASA mission designed to test two effects of general relativity finally came to a close more than fifty years after it was first proposed. The Gravity Probe B satellite was launched in 2004 to test two predictions of Einstein's theory. The first was the familiar phenomenon that massive objects such as Earth warp space-time (known as the geodetic effect). The second was a subtler effect: massive spinning objects, like Earth, drag space-time along with them the way the blades of a blender drag a thick batter as they whirl (known as the frame-dragging effect).

The satellite, which circled Earth in a polar orbit for seventeen months, relied on four spherical gyroscopes, each the size of a ping-pong ball, to make its measurements. A gyroscope's axis of rotation should always point in the same direction if it is isolated from all forces. But geodetic and frame-dragging effects would reorient the axis of each gyroscope ever so slightly.

Researchers at Stanford University had designed Gravity Probe B to measure frame dragging with a precision of 1 percent, but they had to settle for measurements nearly twenty times less accurate. Although the gyroscopes were nearly perfect quartz spheres, their niobium coating trapped electrical charges that made the devices wobble unexpectedly, independent of any relativistic effects.

By the time scientists announced the final results, they had been scooped by a team that had examined the motion of other satellites. In 2004, researchers at the University of Maryland, Baltimore County, had measured frame dragging by monitoring slight shifts in the orientation of two laser-ranging satellites, NASA's Laser Geodynamics Satellite (LAGEOS) 1 and 2. The two satellites slightly shifted the planes in which they orbit. Analyzing those shifts in combination with precision maps of Earth's gravity, the team measured frame dragging with 10 percent accuracy.

Back on Earth, researchers confronted the equivalence principle with quantum theory, replacing heavy weights with atoms. A team from Italy tickled rubidium atoms with a laser, propelling the atoms upward, and then watched as gravity pulled the atoms back down. To probe how quantum theory might affect gravity, the team split the atoms into two clouds with different properties. In one cloud, all the rubidium atoms had a definite, well-defined energy. In the other cloud, the atoms had no definite energy but were described by a quantum state consisting of a superposition of two different energy levels. (The situation is analogous to a qubit, or the quantum version of a computer bit, which can be 0, 1, or a combination of the two. See Chapter 6.)

Each cloud behaves like a light wave, and combining them creates an interference pattern of bright and dark fringes. The team compared the interference pattern to the one generated when both clouds of atoms had the same definite energy. They found no difference in the patterns, indicating that the equivalence principle does not falter when it meets the quantum world.

In another atomic experiment, researchers compared the rise and fall of two different isotopes of rubidium that differ in mass by two neutrons. By merging falling clouds of the two isotopes, the team plans to identify if either isotope accelerates faster than the other.

6

QUANTUM GRAVITY

THEORISTS HAVE BECOME fascinated by the interior of a black hole—a place where extremes meet and the hoped-for union of quantum theory and gravity comes into sharpest relief. Inside a black hole, gravity is immense (the realm of Einstein's theory), and matter appears to crunch down to a small region of space (the subatomic realm of quantum theory). Relativity indicates that a black hole collapses matter to a single point that has infinite density and that space-time itself would disappear, but in that situation Einstein's equation goes berserk. A quantum theory of gravity could rescue Einstein's theory from such catastrophes.

The other fundamental forces in nature—the electromagnetic force between electrically charged particles and the strong and weak nuclear forces that affect particles inside the atomic nucleus—all have a successful quantum theory. But even Einstein, who spent years trying to unify gravity and quantum theory, failed to do so.

In 2009, theorist Mark Van Raamsdonk decided to investigate the murky connection between general relativity and quantum theory. Determined to make the most of his first sabbatical from teaching, he decided to explore an unorthodox approach to the puzzle—one of the biggest in all of physics. After devoting a year to the topic, he submitted a paper to the *Journal of High Energy Physics*.

But in April 2010 the journal sent him a rejection notice, along with a referee's report suggesting that Van Raamsdonk was a crackpot.

Next, he submitted his paper to *General Relativity and Gravitation*, another well-respected journal, but once again he met with harsh criticism. The referee's report was withering, and the journal's editor requested that he rewrite the entire article.

But by then, Van Raamsdonk, who works at the University of British Columbia in Vancouver, was no longer worried. He had entered a shorter version of his paper in the Gravity Research Foundation's annual essay contest, a prestigious competition whose past winners included the renowned theoretical physicist Stephen Hawking. Not only did Van Raamsdonk win first prize, but the award came with a delicious irony: guaranteed publication in one of the journals that had rejected him. *General Relativity and Gravitation* printed the shorter essay in June 2010.

Still, you can't blame the editors and reviewers for being wary. Van Raamsdonk, thirty-five, was trying to marry two of the intellectual tours de force of the twentieth century: Einstein's general theory of relativity, which describes gravity as the curvature of space-time and accounts for the largest structures in the universe, and quantum mechanics, the probabilistic theory that predicts with astonishing accuracy the strange behavior of subatomic particles.

The two theories, both of which predict bizarre phenomena, have long seemed utterly incompatible. Quantum mechanics mandates that a subatomic particle's momentum and position can never be simultaneously known with absolute certainty, and that small particles can be in two places at once. A single electron encountering two adjacent slits doesn't go through one or the other but somehow passes through both.

Einstein's theory of gravity predicts that space-time is as malleable as Jell-O but does not dabble in quantum uncertainties. An electron may go through two slits at once, but relativity allows for no corresponding splitting of the electron's gravitational field.

Unifying the two theories, which scientists say is critical for a deeper understanding of the universe, is a problem that has haunted physicists beginning with Einstein and inspired a slew of unsuccessful—and yes, more than a few crackpot—theories.

The narrative that Van Raamsdonk had woven—that the weirdest property of quantum theory actually gave birth to gravity—may have seemed outlandish. But today scientists say that the pioneering work of Van Raamsdonk and several other scientists appears to be the most promising idea yet for marrying quantum theory and gravity, and that it could forge a deeper understanding of the nature of space-time, information, and the detailed workings of the subatomic world.

In the usual approach to constructing a theory of quantum gravity, physicists would start with Einstein's classical (non-quantum) theory of gravitation and consider how it might be modified to include the statistical nature of quantum theory. But Van Raamsdonk and a cadre of other scientists approached the problem from the other direction: they started with the statistical nature of quantum theory. What they found, to their surprise, is that they could stop right there. Quantum theory, their calculations revealed, *already* encodes the essence of geometry, and by extension Einstein's space-time theory of gravity.

More precisely, Van Raamsdonk and his colleagues propose that space-time as we know it—smooth, connected, continuous—emerges from the very quality of quantum mechanics that Einstein believed would ultimately discredit the theory. That property, known as quantum entanglement, is one of the weirdest concepts in physics. It states that the measurement of one subatomic particle instantaneously determines the state of a partner particle—even if the two reside on opposite sides of the Milky Way.

At first glance, entanglement might not seem so strange. After all, if you own only two gloves, one black and the other white, and you observe your white glove, you know your other glove must be black, even without seeing it. But in the quantum world, nothing is black and white. Each glove has a color that is indeterminate—black, white, and a superposition of the two—until someone performs a measurement and determines the color. Quantum entanglement says a color measurement of one glove not only fixes the color of that glove but at the very same instant determines the color of the other, though the two may lie light-years apart.

Einstein famously derided this state of affairs as "spooky action at a distance," declaring that it would violate the principle that nothing, even information, can travel faster than light. But quantum entanglement, which can be understood as a measure of the information carried by a quantum system, is now an experimentally proven phenomenon that does in fact preserve all of the laws of physics.

The proposed connection between entanglement and geometry depends critically on another strange quantum concept, one that is as profound as it is weird. According to some models, the cosmos as we know it may be, in essence, a hologram: all the action and physical laws in the four-dimensional universe (three

A soup can illustrates the holographic principle. A quantum system with no gravity resides on the can's surface. Another, more complex system, which is subject to both the laws of quantum mechanics and gravity, lies within the volume of the can. The two systems are equivalent: the inner one is a hologram of the outer one *(Courtesy Kristen Dill.)*

dimensions of space and one of time) are governed by a simpler and flatter system that has one less dimension and resides on the boundary of the cosmos. It's as if the label on a Campbell's soup can encoded the contents of the cream of mushroom soup in the container, or if the walls of a theater provided all the details of the three-dimensional drama playing onstage. For Van Raamsdonk the hologram idea was similar to a three-dimensional video game operated by a two-dimensional memory chip. All the three-dimensional information could be read off the two-dimensional chip. The chip and the video game each provide a complete description of the action.

The hologram principle dates to physicist Jacob Bekenstein's early 1970s studies of the physics of black holes. To the surprise of many physicists, Bekenstein calculated that black holes should have entropy—information about a system stored in the variety of possible ways that atoms and other tiny particles interact with one another. But it's quantum mechanics that governs the way those particles interact. That is, if black holes have entropy, they are intrinsically quantum mechanical in nature, even though they are predicted by Einstein's classical theory of gravitation.

Bekenstein's work revealed another surprise. It seemed natural to suppose that if a black hole had entropy, it would be proportional to the number of particles the black hole contained. And that number was related to the black hole's volume: the greater the number of particles, the bigger the volume. But scientists found that the amount of entropy stored within a black hole is revealed not by its volume but by its surface area—in particular, the area of its event horizon, the spherical boundary inside which particles remain forever gravitationally trapped. The larger the event horizon, or area, of a black hole, the larger its entropy.

In the early 1990s, Dutch physicist and Nobel laureate Gerard 't Hooft of Utrecht University in the Netherlands and American string theorist Lenny Susskind of Stanford University went further, proposing what they called the "holographic principle" as a way to understand puzzles about quantum theory, black holes, and the preservation of information. In this view, the area of an event horizon isn't merely proportional to a black hole's entropy; it *is* the entropy. (For more on information and black holes, see "Deeper Dive: Black Holes and the Information Paradox.")

This connection between physics, geometry, and information theory had already inspired John Archibald Wheeler's 1990 exhortation to derive "it from bit." To unify quantum theory with gravity, Wheeler asserted, physicists would have to incorporate new ideas from information theory.

In 1997, physicist Juan Maldacena of the Institute for Advanced Study in Princeton, New Jersey, took a giant leap forward in quantifying the holographic principle. Just as the two-dimensional surface of a hologram contains all the information inside the 3-D illusion, so the "surface" of a holographic universe would contain all the information that we perceive as space-time. Specifically, Maldacena calculated a mathematical equivalence—what physicists call a duality—between a quantum field theory that resides on the surface, or "boundary," of the universe and does *not* contain gravity, and a quantum field theory that describes the volume of the universe, or "bulk," and *does* include gravity.

For Van Raamsdonk, it was a miraculous connection. It meant that if you wanted to describe quantum gravity within a volume, you could do so by examining a simpler, flatter system—an ordinary quantum theory, without any gravity, on the surface enclosing that volume. The particular quantum theory that Maldacena employed on the surface is known as conformal field theory (CFT).

The space-time for which Maldacena showed the connection, known as anti–de Sitter space (AdS), is saddle-shaped, or negatively curved, and neither expands or contracts, unlike the expanding, positively curved geometry of our cosmos. (The model is simpler to work with than the actual universe.) But the finding was still a milestone. Maldacena's duality gave physicists license to think about quantum gravity without thinking about gravity.

At the beginning of his 2009 sabbatical, Van Raamsdonk pored over an article by Maldacena that considered a special type of gravitational system: two black holes connected by a bridge or shortcut called a wormhole. He calculated that the configuration had a holographic double: two separate quantum systems on the boundary that had no gravity but were entangled with each other. Van Raamsdonk wondered if the entanglement of the systems on the boundary was creating the geometric connection between the two black holes within the volume. To test the idea, he tried what theoretical physicists must often resort to: a thought experiment.

He imagined a simpler system than Maldacena's, consisting of quantum fields residing on the surface of a ball. Those fields described empty space-time inside the ball. What would happen if he severed the entanglement between quantum fields located on opposite sides of the ball—if they no longer interacted or were correlated in any way? Making it easier to answer the question was a 2006 paper by Shinsei Ryu and Tadashi Takayanagi, then at the University of California, Santa Barbara. Ryu and Takayanagi had developed a formula that related a specific measure of entanglement on the boundary of a system to a specific geometric quantity in the corresponding space-time. This formula helped to give the quantitative answer to Van Raamsdonk's thought experiment.

When Van Raamsdonk severed the fields on the boundary of the ball, he found it was like cutting in half the memory chip that ran the three-dimensional video game. The empty space-time inside began elongating and pulling apart until, like taffy stretched too far, it was completely torn asunder. When the entanglement vanished, the two regions of connected space-time completely pinched off from each other.

In the absence of entanglement, space-time consists of little dissociated chunks. As theorist Brian Swingle of the University of Maryland in College Park envisioned it, entanglement knits the chunks together to form a smooth space.

It was the ultimate plot twist, Van Raamsdonk thought. Scientists had labored for years trying to figure out how to incorporate quantum mechanics into the study of space-time and gravity. Yet all along, quantum mechanics had contained the ingredients from which emerged space-time—and, by extension, Einstein's geometric theory of gravity.

The idea remains a conjecture, but theoretical physics has seen a veritable explosion of papers on the topic. Other lines of research support the concept. In 2013 Maldacena sent an email to Lenny Susskind that contained a cryptic equation: $ER = EPR$. For Susskind, the message was like getting a shot of adrenaline; he felt like his head would explode. Maldacena was referring to two land-mark papers, both written in 1935. One of them, by Einstein and American-Israeli physicist Nathan Rosen ("ER" for short), showed that general relativity could connect the interiors of two black holes, which appeared to be separate from the outside, by a shortcut through space. The shortcut, called an Einstein-Rosen bridge, is more commonly known as a wormhole. The other paper, written by Einstein, Rosen, and American physicist Boris Podolsky ("EPR" for short), was the first to describe the strange phenom-enon of quantum entanglement.

Maldacena was proposing that the two papers had much more in common than their publication date. Entanglement and worm-holes, he suggested, are two sides of the same coin. If two black holes are entangled, then they are connected by a wormhole; if two black holes are connected by a wormhole, then their outsides are necessarily quantum entangled.

What Maldacena meant, and what he soon described in detail in a paper he coauthored with Susskind, was that entanglement gives birth to wormholes. Maldacena and Susskind conjectured that the relationship between entanglement and wormholes ex-tends beyond black holes. Any time two subatomic particles are entangled, they may be connected by a tiny, quantum version of a wormhole. Wormholes, like other types of space-time curvature, are governed by Einstein's equations of general relativity. So if entanglement gives birth to wormholes, it may give birth to Einstein's geometric theory of gravity as well.

Indeed, researchers including Sean Carroll, of Caltech, and his colleagues have shown that changes in entanglement can lead to changes in the geometry of space-time, and that Einstein's equa-tions of general relativity can even emerge from such changes. Knowing the connection between entanglement and space-time lends insight but still does not provide a complete theory of quantum gravity. The problem is that physicists lack a full dictionary that would allow them to translate a complicated gravitational question in a volume to a tractable problem in quantum theory on the boundary. The goal is to understand quantum gravity by reformulating interesting gravitational questions in the language of field theory, which physicists understand well. But it isn't always clear how to do the translation. For instance, no one yet has a full representation on the boundary for what happens behind the event horizon of a black hole in the volume.

To help develop a better dictionary that would translate between gravity and quantum theory, physicists in the 2010s began embracing ideas from two other disciplines in which entanglement already plays a starring role—the fledgling field of quantum computing and the study of how particles in solids behave.

A quantum computer aims to outperform a standard computer by exploiting the counterintuitive properties of quantum physics. Standard computers store information as bits, which are restricted to just two values—0 and 1. In contrast, information in the quantum world is encoded by qubits, pairs of quantum states that can have the value of 1, 0, or a superposition of the two.

In principle, that rainbow of possible states—if properly entangled with other qubits—is what would enable a quantum computer to perform calculations an ordinary computer could never finish, even if it had the entire history of the universe in which to do so. But that ability depends on preserving the fragile entanglement among the qubits. Once these correlations between qubits are destroyed—and even a slight disturbance from the outside world can inadvertently do so—the qubits "collapse" to either 1s or 0s, and quantum computations are no longer possible.

That fragility was once deemed a fatal flaw in trying to construct a quantum computer. But in 1995 physicists were surprised to find the existence of so-called quantum error-correcting codes—strategies that could repair broken or corrupted correlations between qubits and allow physicists, in theory at least, to take full advantage of the power of a quantum computer.

Error-correcting codes repair broken correlations by building in redundancy in the way a qubit (or a bit) is labeled. Consider an example from the non-quantum world: Let's rename the standard bit 0 as "000" and the standard bit 1 as "111." Imagine these bits became corrupted, with a portion of each label no longer readable. Even if two numbers got lopped off the label "111" so that it read "1," the bit could still be correctly identified.

Quantum error-correcting codes work in a similar manner. One hallmark of any type of error-correcting code is that the encoded information should be nonlocal—that is, spread out over a large region of space. Without such widespread distribution, damage to a single spot could prevent recovery of the original information and the error-correcting code would fail.

If a set of information is likened to the pages of a book, quantum error-correcting codes show that the information is not stored within any one particular page but is spread out in the correlations *between* the pages—how one page relates to another. And if a few adjacent pages are torn out, the information may still be accessible, because the widely distributed correlations may be preserved.

The curious thing, which has excited quantum information specialists, is that Maldacena's correspondence between the bulk and the boundary embraces a similar kind of nonlocality. Information in the bulk can be directly mapped to information on the boundary, but the mapping is highly nonlocal—any one qubit, or entangled pair, in the bulk corresponds to many widely spaced qubits, or entangled pairs, in the boundary. The closer a qubit lies to the center of the bulk universe, the more boundary qubits are needed to reconstruct the information.

The findings suggest that once entanglement knits space-time together, it may be difficult to tear it apart. A few disruptions or defects in entanglement on the boundary won't unravel the fabric of space-time.

The relationship between quantum error-correcting codes and Maldacena's duality may go beyond mere similarity. A group of physicists including Daniel Harlow, of MIT, and John Preskill, of Caltech, has suggested that the duality is a direct example of a quantum error-correcting code. In fact, Harlow and colleagues have demonstrated that the analogy is a precise one for simple models. By identifying the holographic duality with quantum error-correcting codes, Harlow hopes to develop new translations between the bulk and the boundary. If they hit upon the right translation, it might answer two long-standing questions: What is the physics inside a black hole? And does information captured by a black hole ever really come back out?

Another route to strengthening the translation between the bulk and the boundary has come from scientists who study the physics of solid materials. Entanglement became an essential ingredient of such studies in the 1990s. Physicists analyze the complicated interactions of billions of electrons within a solid material to understand and predict the material's basic properties, such as how well it conducts electricity or transmits light.

But even keeping track of all the possible interactions of a tiny group of, say, 300 electrons can be a daunting task. Electrons possess a quantum property called spin, which points either up or down. To specify the quantum state of the entire system, which must include the probability of every possible configuration of the 300 spins—such as electron no. 1 with spin up, the rest down, electron no. 1 with spin down, the rest down, and so on—would require 2^{300} possibilities. Writing down the probability of each of those myriad configurations involves more information than can be stored in the entire universe.

To approximate such a complicated quantum state, physicists who study condensed matter use mathematical shorthand called a tensor network. Tensors keep track of objects that have more than one number or property associated with them. (The simplest kind of tensor, a vector, tracks two properties. A velocity vector, for instance, indicates how fast someone is running and the direction in which she is moving.)

Consider a line of interacting atoms. A tensor network starts out simply, initially accounting for only the entanglement between neighboring particles, which are the ones most likely to interact. A tensor connects these pairs of particles the way a Lego building block bridges two adjacent Lego pieces. There are many of these small Lego building blocks (one for each connected pair), and each becomes a node for considering the entanglement between groups of atoms that lie farther apart on the same line. The net effect is to create a hierarchical, geometric pattern that resembles a Christmas tree.

That's how Brian Swingle was using tensor networks in 2007. Bearded and burly, Swingle was a graduate student at MIT, studying electron interactions in solids, when he decided to take a course in a branch of quantum theory called string theory. When his teacher explained the bulk/boundary work of Maldacena, Swingle noticed an intriguing pattern. The mathematical machinery that translates a universe on the boundary to the corresponding universe in the bulk bore an uncanny visual resemblance to a special kind of tensor network Swingle and others employed to track the quantum states of electrons in solid materials.

Swingle wasn't sure how far the correspondence would go. But he soon proposed that in building up entanglement, a particular tensor network known as MERA (multiscale entanglement renormalization ansatz) builds up space-time. The tensor network could be thought of as creating an extra dimension that can be interpreted as the emergence of geometry in a higher-dimensional space, the bulk. He first published the link between MERA and AdS/CFT in 2009, a year before Van Raamsdonk's work on entanglement.

Anti–de Sitter space, according to Swingle's calculations, is both a tensor network representing the entangled state of a boundary system and a space-time in its own right. Using tensor networks, Swingle, Van Raamsdonk, and their colleagues have shown that changes in entanglement on the boundary reproduce a simplified version of Einstein's equations of general relativity (his classical geometric laws of gravitation) in the bulk.

Yet as enamored as many theoretical physicists have become of entanglement as a route to developing a theory of quantum gravity, it can't be the whole story—a sentiment Susskind succinctly captured in the title of a 2016 paper: "Entanglement Is Not Enough." Entanglement has enabled researchers to recover Einstein's equations in the bulk and examine quantum gravity in systems where gravity is relatively weak, Susskind notes. But by itself this strange quantum property has not allowed physicists to peek inside a black hole. That's where powerful gravitational fields crunch matter into a vanishingly small space—the very realm in which quantum gravity must rule.

Another drawback of entanglement is that it reveals the interaction between particles only at a given moment, a snapshot frozen in time. To find a complete quantum gravity theory, time must be taken into account.

For that, Susskind asserts, physicists may have to borrow another concept from information theory: computational complexity. The term refers to a way of quantifying the degree of difficulty in carrying out a task. In particular, computational complexity is the minimum number of operations required to complete a task—whether it's the minimum number of steps required to walk from the North Pole to the South Pole, the minimum number of logical operations needed to build a computer program, or the minimum number of mathematical steps necessary to fully describe a quantum system.

Because a quantum system has a wider range of possible configurations than a classical (non-quantum) system, it has the capacity to have much greater computational complexity. In fact, the complexity of a quantum system increases dramatically with time, until it reaches a maximum value.

Susskind began to think about complexity several years ago when he realized that one solution to Einstein's equations of general relativity showed that the interior of a black hole—the volume of space behind its event horizon—elongates with time. That posed a puzzle: If that was what was going on in the bulk, what corresponding property might be varying on the boundary? Susskind knew it couldn't be entanglement, because such correlations between particles reach their maximum in less than a second and can't get any bigger.

Only one quantity was left—the internal structure of the quantum system on the boundary. He and physicist Douglas Stanford, of the Institute for Advanced Study, examined their black hole model in detail and found that its complexity increases with time in the same way that computational complexity would.

The take-home message is that if quantum entanglement glues space-time together, then computational complexity drives the growth of space-time, at least in the case of black holes. Computational complexity, Susskind suggests, might even play a role in the growth of space-time in our universe, which is expanding at a faster and faster rate. And because complexity seems very much linked to the activity inside a black hole, it may provide additional insight into how to formulate a complete theory of quantum gravity. But, says Susskind, the role of complexity is a sign that physicists will have to reach beyond entanglement and the holographic principle in order to develop that theory.

In science, however, theory is one thing, observation quite another. Could scientists actually find a way to observe the evidence for general relativity's wildest predictions? As it turned out, they would hear the evidence before they would see it.

....
....

DEEPER DIVE: Black Holes and the Information Paradox

Can the cosmos keep a secret?

Suppose there's embarrassing information you want destroyed—the notation in your diary about your girlfriend's sister, that cellphone video of you doing the Macarena, or the poster of One Direction on which you wrote "I Love Niall" in pink crayon.

You could shred the paper evidence and erase the video, but a tenacious detective could reassemble the shreds, and a computer geek could restore the video. You could burn the material, but a forensic scientist could use the ash, carbon dioxide, and other combustion by-products to reconstruct every item. Desperate, you scour the universe and find the ultimate solution: drop the material into a black hole. Swallowed by this gravitational trap door, in which everything falls in but nothing gets out, surely your secret will finally be safe.

Or will it? Physicists have spent the better part of three decades arguing about whether information can ever be retrieved from a black hole. The original debate revealed a dramatic clash between two highly successful theories about the universe—general relativity and quantum theory. Quantum theory demands that information always be preserved and accessible, but general relativity seemingly allows for data to vanish from view.

Scientists love such paradoxes because they have led to startling new advances in physics, including the birth of quantum theory. Solving the black hole information paradox not only would reconcile quantum theory with general relativity but may also lead to a new understanding of space-time and the universe that could rival Einstein's achievement. In fact, physicists have even bet on it. In 1997, Stephen Hawking and cosmologist Kip Thorne, of the California Institute of Technology, bet another Caltech theorist, John Preskill, that if a book fell into a hole, the information in that volume would never reemerge.

Information that fell into a black hole would be impossible to retrieve as long as the black hole existed. But Hawking himself knew that black holes don't live forever. In 1974, using elements of quantum theory, he calculated that black holes leak radiation.

The radiation arises because the vacuum in quantum theory isn't empty. It's a cauldron of particle-antiparticle pairs popping in and out of existence. If such pairs come into existence just outside the event horizon, sometimes one particle will fall Into the hole, while the other will remain outside. The outside particle escapes the black hole's tug and streams freely into space, Hawking realized. From the point of view of an outside observer, the black hole radiates energy, a phenomenon known as Hawking radiation.

Because it is losing energy, the black hole gradually reduces its mass and eventually winks out of existence. How fast the black hole evaporates depends on its size: the bigger it is, the more slowly it evaporates. A black hole the size of the Sun would fade away in about 10^{66} years. That's a long time, but it isn't an eternity. So it would seem that information *does* emerge from a black hole, carried by the radiation.

But Hawking said no. He calculated that the information carried by the leaking radiation would be so scrambled as to be nonsensical—a Humpty Dumpty that not only couldn't be put back together again but couldn't even be identified as having once been an egg.

If a dictionary fell into a black hole, the emerging radiation would not contain a single definition. All the data would be lost.

That might be comforting for someone trying to keep a secret, but it's anathema to a core principle of quantum theory: that the present always preserves information about the past. Quantum theory describes a subatomic system in terms of probabilities—for instance, how likely it is that the measurement of the energy of an electron will yield a particular result. Even though quantum theory deals with likelihoods rather than certainties, the theory demands that a subatomic system evolve in a predictable fashion. Given the state of a quantum system at a particular instant in time, an observer can always determine the state at an earlier or later time. But that can't happen if information is lost or irretrievable.

In 2004, Hawking decided he was wrong after all and conceded the bet. He revised his thinking based in part on an emerging concept known as complementarity. Developed by Lenny Susskind, complementarity argues that information can be in two places at once.

On the one hand, it can cross the event horizon, the one-way boundary between a black hole and the rest of the universe, and fall inside. On the other hand, it can also remain smeared out on the event horizon and not fall in. This bizarre behavior is perfectly permissible because no single observer can reside both inside and outside a black hole at the same time. An observer sees only one copy of the information. In this scenario, information is not lost.

In 1997, Juan Maldacena extended the notion of complementarity by applying it to anti–de Sitter (AdS) space. In three dimensions, AdS space can resemble a cylinder. Maldacena considered a system inside the cylinder that obeys both the laws of quantum mechanics and gravity. He showed that this model universe is exactly equivalent to a simpler system on the cylinder's surface, one in which gravity does not exist and only the rules of quantum mechanics apply. If Maldacena was right, information had to be preserved, because it remained on the surface.

"We have to reconcile the fact that information is pasted on the horizon with the fact that the infalling observer's experiences are consistent with nothing special at the horizon," says Susskind.

Despite this new understanding, Kip Thorne, unlike Hawking, never did concede the bet. That may have been just as well, because in 2012 physicists uncovered another twist in the information paradox. Theorists Joseph Polchinski, Ahmed Almheiri, Donald Marolf, and James Sully considered what would happen to a pair of quantum particles just outside a black hole that were entangled, or correlated, with each other.

If one of the entangled partners fell into the black hole while the other remained outside, the two particles would remain entangled. But the outside particle would also be entangled with other outside particles that had previously streamed into space as Hawking radiation. That can't happen in quantum theory, which forbids a particle to be simultaneously entangled with two separate sets of particles—in this case, those inside a black hole and those outside it.

To remedy the situation, Polchinski and his colleagues performed a thought experiment: they tried severing the entanglement between the particles on either side of the event horizon. When they did so, they found that a wall of energy formed at the horizon, a shock wave that incinerated anything trying to cross the boundary. The so-called firewall solved the quantum entanglement issue, but it created a big problem for general relativity.

Einstein's principle of equivalence in its strongest form states that someone freely falling in a gravitational field experiences the same laws of physics as someone freely floating in a gravity-free environment. The principle holds even for someone in free fall just outside a black hole. But if that's true, someone passing through the horizon can't be incinerated. In burning up—something that would happen only in the vicinity of the black hole and not elsewhere—the laws of physics would not be the same as for someone floating in a gravity-free region of space. The principle of equivalence would be violated. In effect, Einstein's theory of general relativity carries a "no-drama" clause: nothing special should happen as someone crosses the event horizon.

In 2013, Susskind and Maldacena posited that the firewall was entirely unnecessary. Their theoretical work suggested that the entanglement between particles creates wormholes, or tunnels, between widely separated regions of space. The tunnel would directly link particles trapped inside the black hole to those that had long ago left the black hole and streamed into space as Hawking radiation. The connecting tunnel upholds quantum theory's rule that a single particle can't be simultaneously entangled with two separate groups of particles. The particles inside and outside the black hole are not separate but connected through the tunnel.

There's another argument against the firewall scenario. Daniel Harlow, a quantum physicist at Harvard University, and Patrick Hayden, a physicist and computer scientist at Stanford University, considered whether someone could ever detect the information both inside and outside a black hole. To do so, the observer would have to decode the information contained in the Hawking radiation, then dive into the black hole to examine the data contained in the infalling particles. The researchers calculated that deciphering the Hawking radiation would take such a long time that the black hole would evaporate before the observer was ready to dive in.

Most physicists now believe that information does emerge from black holes. But they still have to show exactly how it happens. Hawking found that his namesake radiation would be too scrambled to transmit information, and no one has been able to perform a calculation that would refute that assertion.

There may be another way out of the black hole information paradox, and it involves an exotic type of wormhole. Wormholes, of course, are already unusual beasts. These tunnels between two distant regions of space-time connect two black holes that otherwise would have no link. In many theoretical models, a wormhole collapses before any material or information can pass through. But in 2017, building upon earlier work by Maldacena, three researchers found that if two black holes connected by a wormhole are quantum mechanically linked in precisely the right way, the throat of the wormhole stays open and information travels through it. Though physicists have just begun exploring this possibility, it could be a way to recover information from the scrambled Hawking radiation.

In the meantime, someone trying to find the ultimate hiding place for an embarrassing Macarena video might have to keep searching.

7

HEARING BLACK HOLES

THE MOST STUNNING confirmation of Einstein's theory of general relativity came from observations of a black hole—actually the merger of two of them. The discovery announced itself with a fleeting signal that arrived on Earth in mid-September 2015.

When MIT physicist Scott Hughes first saw the image of that signal on a colleague's cellphone, he felt a rush of emotion he had experienced only twice before—when he saw his newborn daughter's face for the first time and his dying father's face for the last time. His colleague kept talking, but Hughes couldn't hear him. All he could think about was the image on the cellphone. It showed a pattern of wiggles that first grew in amplitude and increased in frequency, and then rapidly diminished in amplitude—a picture that Hughes had been seeing in his mind's eye since the dawn of his career more than twenty years earlier.

In 1995, Hughes was a graduate student at the California Institute of Technology, studying how black holes, the powerful gravitational traps from which not even light can escape, affect their surroundings. Hughes knew that a black hole at rest would dent space-time the way a bowling ball would sag a rubber sheet. But what would happen to space-time if heavy objects were shaken or accelerated, like two black holes about to smash into each other?

Just as a bouncing bowling ball would jiggle the rubber sheet, shaking a chunk of matter would generate undulations in the fabric of space-time. These undulations, known as gravitational waves, would spread out across the universe like ripples in a cosmic pond.

As a wave passes, it distorts the distance between two freely suspended masses in a particular pattern. Along the direction the wave is traveling, it has no effect on the distance. But perpendicular to that direction, the wave stretches the distance along one dimension while shrinking it along the other. Half a cycle later, the undulating wave deforms space in the opposite sense, shrinking the distance along the dimension it had stretched and stretching the distance along the dimension it had shrunk.

Gravitational waves had never been found (see "Deeper Dive: Gravitational Waves Lost and Found"), but Hughes had become captivated by a new, more sensitive gravitational-wave detector that was about to come online. He and a colleague calculated that an advanced version of the detector, known as LIGO (Laser Interferometer Gravitational-Wave Observatory), would have the best chance of sensing the waves if each of the merging black holes had a mass between twenty and sixty times that of the Sun.

At the time Hughes and his collaborator did their calculation, researchers lacked the computational tools to determine what the gravitational waves would actually look like—their size, duration, and change in frequency. That breakthrough came nearly a decade later, when other researchers developed the techniques to solve Einstein's equations of gravity on computers.

Simulation of two black holes about to merge. The first gravitational wave ever recorded on Earth came from such a merger and was one of the most striking confirmations of Einstein's general theory of relativity. *(SXS Project/NASA.)*

The match between those computer calculations and the trace from LIGO that Hughes was now seeing on his colleague's cellphone was more than uncanny. It was exact. A detailed analysis would later show that the masses of the two coalescing black holes were smack in the middle of the range that he and his collaborator had calculated nearly twenty years earlier.

After more than five decades of hunting for these space-time ripples, the very first gravitational waves ever discovered had been generated by the collision of two stellar-mass black holes, just as Hughes had envisioned. And it took LIGO's twin installations, each of which records changes in length smaller than one ten-thousandth the diameter of a proton, to find it (see "Deeper Dive: LIGO and Beyond").

Although LIGO is called an observatory, it lacks a telescope, dish-shaped mirrors, or other light-gathering accouterments to make sharp images of the heavens. The light seen by our eyes is but a tiny sliver of what's known as the electromagnetic spectrum, which includes gamma rays, X-rays, ultraviolet, visible light, infrared radiation, and radio waves. All these types of light can produce fine-grained portraits of the cosmos because their wavelengths are typically much smaller than the planet, star, or galaxy emitting the radiation. And because light interacts strongly with matter, lenses can bring this radiation to a focus.

Gravitational waves, in contrast, are generated by the large-scale motion of massive bodies and typically have wavelengths much larger than the objects generating them. The long wavelengths mean that gravitational waves can't be used to image a celestial body or reveal its shape. And because the waves interact only weakly with matter, they aren't easily brought to a focus the way light waves can be. But if gravitational waves can't picture the universe, they can, in effect, provide the soundtrack. Just as light-collecting telescopes gather electromagnetic radiation to view the universe, LIGO and other gravitational wave detectors act as listening posts, picking out the space-time vibrations from spurious noise.

Although they can't be heard, gravitational waves have several properties in common with sound. Sound waves generate acoustic signals—a bang, a shout, or a Mozart concerto—by alternately stretching and compressing the medium, such as air or water, through which they travel. Gravitational waves generate vibrations by alternately stretching and compressing a material—the fabric of space-time.

Converting gravitational waves into sound, says Hughes, highlights the rich information that the frequency, amplitude, and duration of the waves carry about the systems that created them. From these properties, scientists can determine the mass, density, and rate of rotation of the source. It also happens that the frequency range of the gravitational waves that LIGO can detect, 10 to 1,000 cycles per second, falls within the audible range.

The first gravitational waves recorded on Earth were forged long ago in a galaxy far, far away when two widely separated black holes ever so slowly spiraled toward each other. Even during the earliest days of this fatal attraction, when the black hole partners moved around each other in a genteel pas de deux, the duo shook space-time. But those initial vibrations, akin to a steady whisper, were too faint and too low in frequency for LIGO's detectors.

Over thousands to millions of years, as the black holes inched ever closer, their slow dance became a furious death spiral. Only during the final two-tenths of a second before the black holes collided, as the space-time ripples grew stronger and rose higher in frequency, could LIGO hear them. Translated into audio, the waves resembled the chirp of a bird gliding up the musical scale.

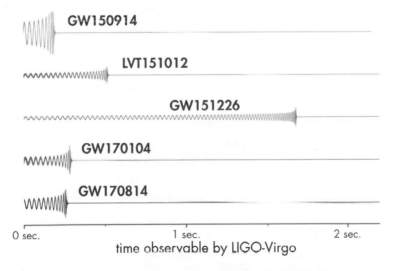

Duration of some of the first "chirps"—gravitational wave signals—recorded by LIGO and Virgo. The signal at the top, GW150914, which heralded the merger of two black holes, arrived at LIGO's two detectors on September 14, 2015, and was the first gravitational wave ever recorded. *(Courtesy LIGO / University of Oregon / Ben Farr.)*

And then the coda: a single note, like the thud of a struck gong, which rapidly died out. The two black holes had merged to give birth to a single, more massive gravitational beast. This part of the atonal symphony LIGO also recorded.

Before the black holes merged, they weighed in at twenty-nine and thirty-six times the mass of the Sun. Although the combined mass of the two black holes adds up to sixty-five Suns, the black hole created by the merger only weighed as much as sixty-two Suns. In a fraction of a second, the other three solar masses were converted into pure energy—the energy carried by the final set of gravitational waves from the collision. The energy carried by the waves, were it available as some mammoth cosmic battery, would have been enough to light up all the stars in the universe's 175 billion galaxies as big as the Milky Way, Hughes estimates.

When the gravitational waves left the distant galaxy 1.3 billion years ago, plant life had just taken root on Earth. By the time the waves reached Earth at 4:50 A.M. Central Time on September 14, 2015, two exquisitely sensitive listening posts had just begun operation. LIGO's twin detectors are spaced 1,865 miles apart, one in Hanford, Washington, the other in Livingston, Louisiana. The waves that washed over them were beyond weak—so feeble that they would deform a four-kilometer-long pipe by only one-thousandth the width of a proton.

When scientists first saw the LIGO signal—the waveform of an unusually loud chirp—it seemed too good to be true. The researchers had good reason to be skeptical. Engineers had only just completed a major upgrade to the detectors. Moreover, the LIGO staff knew that a small cadre of LIGO scientists randomly injected fake signals into the data to keep the LIGO team on its toes. In the past, scientists had spent months studying what had seemed to be a bona fide gravitational wave signal, only to find out, just before they were ready to publish their results, that it wasn't real.

But this time there had been no injection. And both the Hanover and Livingston detectors had caught the same wave, seven milliseconds apart.

The discovery gave striking confirmation to Einstein's geometric theory of gravity. In 1916, Einstein predicted that when a massive object explodes, crashes into another object, or otherwise speeds up or slows down, it generates vibrations that shake spacetime as if it were a bowl of Jell-O. Detection of the quakes not only validated Einstein's theory of gravity but confirmed in a radically new way that black holes actually exist. Previous evidence—and there was plenty of it—was indirect. Astronomers had clocked the unaccountably high speeds of stars whipping around the dense cores of galaxies and recorded the energetic radiation from the hot disk of gases orbiting just outside suspected black holes. LIGO's evidence was different. The gravitational waves the observatory recorded came from the colliding black holes themselves. A culmination of forty years of work on detector technology that can sense changes in distance smaller than the nucleus of an atom, the discovery won the 2017 Nobel Prize for veteran LIGO physicists Rainer Weiss, Kip Thorne, and Barry Barish.

Some on the LIGO team have likened the import of the discovery to the first time light was captured on a photograph: a snapshot of his Burgundy, France, estate that Joseph-Nicéphore Niépce took from an upstairs window in 1826 or 1827. But a more apt analogy may be the first time Thomas Edison captured sound on a phonograph: his own voice reciting "Mary Had a Little Lamb" recorded on a piece of tinfoil in 1877. "It sings! It laughs!" exclaimed posters publicizing Edison's phonograph.

Years before LIGO detected gravitational waves, scientists had used the world's finest optics to image massive stars and record the light from material swirling around giant black holes in distant galaxies. But now scientists had their ears open to the universe as well. They had heard the song of two black holes merging. They had heard gravity for the first time.

Over the next twenty-three months, LIGO detected gravitational waves from the collision of several more pairs of black holes. But for astronomers, LIGO's sixth detection, in collaboration with Virgo, its Italian-based counterpart, proved golden. For the first time, astronomers and physicists simultaneously recorded gravitational waves and light from the same stellar explosion. The gravitational waves identified the participants in the explosion and cued the visual observations, which revealed a fiery display of cosmic alchemy—a cauldron that seeded the universe with precious metals.

The gravitational waves that LIGO and Virgo recorded were once again generated by the collision of two partners in a remote galaxy. But this time the duo were neutron stars—the shrunken, superdense remains of massive stars that had explosively cast off their outer layers. A teaspoon of the stuff would weigh more than Mount Everest.

Unlike black holes—the Roach Motels of the cosmos, from which everything goes in and nothing goes out—colliding neutron stars send all kinds of fireworks into space, from high-energy gamma rays to visible light and radio waves. As with the black hole partners, the neutron stars became locked in a gravitational embrace while still at a respectable distance from each other. The initial gravitational waves they generated were a mere whisper that barely rippled the surrounding space-time pond. Over time, the stars cinched in, orbiting faster and faster until in the last frenzied moments they whipped around each other at close to the speed of light. The whisper had grown to a piercing trill as the neutron stars smashed together. And at 8:41 A.M. Eastern Daylight Time on August 17, 2017, LIGO and Virgo recorded that trill.

A mere 1.7 seconds after the gravitational waves reached Earth came a light show. NASA'S Fermi Gamma-ray Space Telescope, which orbits Earth to record some of the highest-energy radiation in the universe, had detected a short-lived burst of gamma rays. The explosive collision of two neutron stars to form a black hole is believed to have unleashed equal and oppositely directed jets of gamma rays into space.

At the time, no one knew if the gravitational waves and the gamma-ray burst were related. By itself, LIGO can't pinpoint the location of the waves it detects to anything better than 600 square degrees, or about 3,000 times the area of the full moon on the sky. But new information changed that. Scientists working with the Virgo gravitational wave instrument outside Pisa, which had come online only two weeks earlier, found that their detector had recorded a tiny signal from the same wave. The Virgo signal was so faint that it had barely registered, indicating that the wave had passed close to one of the detector's few blind spots. That information proved to be the key to localizing the cosmic event to a relatively small swath, 28 square degrees, in the southern sky. Within hours, the LIGO-Virgo collaboration alerted astronomers to look for a visible counterpart to the space-time jiggle in the same part of the sky. Gamma rays from the heavens can't be detected on Earth because our planet's atmosphere absorbs them. But ground-based telescopes can detect the afterglow from a gamma-ray burst—lower-energy radiation ranging from visible light to radio waves. Astronomers around the world scrambled to be the first to detect the afterglow. Over the next week, more than 70 telescopes on all seven continents would follow the light show.

Ryan Foley, a young astronomer at the University of California, Santa Cruz, was in Tivoli Gardens, taking his first break from a month-long summer school on gravitational waves at the University of Copenhagen, when he received an urgent text message from Dave Coulter, one of his graduate students also at the summer school: "Stop what you're doing and check your email."

In line with his partner to ride the amusement park's ornate carousel, Foley could not check his messages, so Coulter filled him in about the gravitational wave signal and the gamma-ray burst. Foley thought it might be a prank, and replied: "I might leave [the park] but if you're joking and don't tell me now I will not be amused."

Coulter texted back: "I'm not joking. Jesus, man, I wouldn't joke about this."

In a daze, Foley boarded the ride, his girlfriend guiding him to one of the wooden circus animals—an elephant or a giraffe, he couldn't remember. As the carousel turned, his mind was in a whirl.

When the ride finished, Foley raced on his bicycle to the university, where his students had already gathered. Foley contacted colleagues on the one-meter Swope Telescope in Chile, where it was still morning, asking them if they would survey the southern sky that night to look for a visible-light counterpart to the burst.

As the images were recorded in Chile, Foley's postdoctoral fellow, Charlie Kilpatrick, compared each exposure to archival images of the same patch of sky, looking for a bright spot that had not been there before. On the ninth exposure he found it: a blue dot in a galaxy called NGC 4993. Kilpatrick appeared to be the first to see it. The team had beaten everyone to the punch.

Over the next few nights the blue dot turned red, several observers found. The switch in color dovetailed with a model of neutron star collisions proposed by Brian Metzger of Columbia University. In his theory, the merging neutron stars hurl debris into space in stages. During their final orbits, the stars would cast out neutron-rich matter from their outer layers, creating a rapidly expanding fireball. Neutrons and the smattering of protons in this debris cloud would bind together to forge heavy elements. The chemical composition of this cloud accounted for its initial blue color.

Not long after, another source of debris, most likely flung from a fat doughnut of material around the collapsed body forged by the merger, struck the glowing cloud, kept hot by radioactivity. Over just a few days, this cauldron created enough gold dust to make about two-hundred solid-gold Earths, turning the fireball red.

Observations with the Gemini South Telescope and the European Southern Observatory's Very Large Telescope, both in Chile, and the Hubble Space Telescope confirmed the fingerprints of precious metals in the debris cloud. The discovery fills in a missing link in the history of cosmic alchemy, explaining how the early universe of hydrogen and helium transformed into a cosmos of planets, stars, and galaxies that contain a plethora of heavy elements.

Astronomers have long known that stellar explosions called supernovas could account for forging elements as heavy as iron, but not for producing heavier elements such as gold, platinum, and uranium. Scientists had theorized that the explosive collision of neutron stars, called a kilonova, provided the pathway, but no one had ever seen one.

The key to the observation was the gravitational wave messenger. It's unlikely that anyone would have followed up on the gamma-ray burst, which was run-of-the-mill and relatively low-energy, had it not been linked with a space-time vibration. And for the first time, astronomers could record a minuscule space-time quake from somewhere in the cosmos, point a telescope into deep space, and find the galaxy it came from. The feat means that astronomers now have a chance to catch exploding stars at the earliest moments of their blow-up, when so much of the activity happens. So because of gravitational waves, astronomers are observing phenomena no one has ever seen before.

As if revealing the origin of the universe's supply of gold and other heavy elements wasn't enough, the gravitational signals recorded on August 17, 2017, opened up two other cosmic vistas.

Celestial events observed both as gravitational waves and as light signals provide a new and more accurate way to measure the expansion of the universe, which has been increasing its girth ever since the Big Bang.

To quantify cosmic expansion, scientists need to measure two numbers: how fast an object appears to be receding from Earth and its distance. Getting the recession velocity is relatively easy, obtained by observing the light and determining its redshift—the extent to which the emitted wavelengths have been shifted to the longer, or redder, end of the electromagnetic spectrum. Obtaining the distance to the object is more challenging.

In the absence of gravitational waves, astronomers relied on "standard candles"—stars or stellar explosions believed to have a known intrinsic brightness. Scientists then compared the known brightness to how bright the objects appeared when observed from Earth. The dimmer the body appears in the sky, the greater its distance. However, stars can vary their brightness for all sorts of reasons, and estimates of the intrinsic brightness are prone to errors.

Gravitational waves don't provide a measure of the brightness of objects, but they do something just as good: they serve as standard sirens. When two massive bodies collide, the frequency of the resulting gravitational waves and the rate at which the frequency changes—the characteristic chirp—suffice to determine the intrinsic strength of the waves (the strength that would be measured by an observer right next to the collision). And, as with the standard candle method, by comparing the intrinsic strength of the waves with the much weaker strength recorded by detectors on Earth, scientists could precisely determine the distance to the merger.

Back in 1986, when physicist Bernard Schutz published that calculation, no one knew if merging neutron stars would create an explosive light show as they merged, or if they would quietly coalesce into a black hole without any light emission at all. Thirty-one years later, in mid-August 2017, Schutz, based at the University of Cardiff in Wales, had just celebrated his seventy-first birthday and had not checked the LIGO archive for several days. When he did, he was astounded. The August 17 event came from a neutron star merger a relatively close 130 million light-years away—close enough to provide an unusually clear and powerful standard siren. "We had no right to be so lucky" the first time around, Schutz says. "This was a real gem."

Though it was only a single event, the signal, combined with observations of the light from the merger, provided a rough measure of the universe's current expansion rate. As the gravitational wave detectors record more sirens, astronomers may be able to pinpoint the expansion rate with an accuracy of better than 1 percent, helping settle an ongoing debate about the rate of expansion using the standard candle method. And by using space-based interferometers that can search for more distant mergers that occurred further back in time, researchers hope to understand more about how and why the universe accelerated its rate of expansion some 5 billion years ago and the nature of the mysterious entity—dubbed dark energy, for want of a better description—that switched on the acceleration.

The gravitational waves generated by merging neutron stars offer a new way to peek inside these compact bodies. The properties of the waves reveal the extent to which neutron stars deform before they coalesce, which in turn depends on their density and compressibility. Although neutron stars are the densest known reservoirs of matter in the universe that haven't collapsed to form black holes, physicists aren't certain what happens inside these stars or exactly what their core density is.

Some theorists suggest that a neutron star consists almost entirely of neutrons and a smattering of protons and other particles, all held at densities two to three times the normal density of an atomic nucleus. Other scientists propose that under the enormous pressures inside a neutron star, the subatomic particles known as up and down quarks, which bind together to make protons and neutrons, break free. The liberated up and down quarks could form a new state of quark matter, turning a neutron star into a quark star. Yet other researchers suggest that other types of quarks, including the strange quark, are created and liberated, forming an even more exotic kind of quark star.

Pinning down the diameter of a neutron star and its compressibility could help distinguish between these theories, because each model provides different estimates for the "squishiness" of the matter that makes up the star. For instance, if the matter is relatively squishy, and therefore easy to compress, a neutron star of a given mass will have a smaller diameter than if the matter is stiff and resists compression.

The final vibrations of the 100-second-long signal from the August 17, 2017, event were too high in pitch for LIGO and Virgo to detect. That prevented the observatories from studying the neutron stars just before they coalesced, when their super-strong gravitational fields would have deformed each other and provided vital information on the stars' compressibility. Even so, researchers were able to determine that the stars were no bigger than 30 kilometers across, a number that agrees with other measurements that suggest the neutron star matter is relatively compressible. Future detections should place stricter limits on the size.

And what exactly did the neutron stars form when they coalesced? Data gathered by LIGO revealed that the object is about

2.7 times the mass of the Sun. That makes it either the lowest-mass black hole ever identified or the most massive neutron star seen to date. LIGO researchers are betting on the black hole scenario: the merged star underwent a collapse so catastrophic that it sealed itself off from the rest of the universe.

If the colliding neutron stars had formed a single, heavier neutron star, the body would have likely produced a torrent of X-rays. But NASA's Chandra X-ray Observatory found only low levels of that type of radiation. Although astronomers have long suspected that neutron star mergers could form a black hole, until now they had lacked strong evidence. The chirp that LIGO and Virgo heard from the August 17 event could indeed have been the birth cry of a black hole.

But if scientists could now, in effect, hear a black hole, would they ever see one? Pointing a network of radio telescopes with the effective diameter of Earth at the center of the Milky Way, astronomers are poised to take the first image of the most extreme example of Einstein's general theory.

....

DEEPER DIVE: LIGO and Beyond

Near a loblolly pine forest in the small bayou town of Livingston, Louisiana, about forty miles east of Baton Rouge and just five miles from Fireworks Warehouse USA, lies one of the quietest places on Earth.

The silent setting is a prerequisite for detecting gravitational waves, weak vibrations that can easily be swamped by stray sound waves, seismic activity, or even the tiny oscillations induced by laser light.

The setup features twin arms, both exactly four kilometers long and set at right angles to each other, forming a giant L that stretches to the horizon. To eliminate sound waves, all the air is pumped out of each tube-shaped arm. Mirrors that lie at either end of the tubes hang from silica fibers to minimize vibration; concrete sheathing protects the chambers from the weather, and nearby vehicles are forbidden from going faster than 10 miles per hour.

That's just some of what it takes for the Advanced Laser Interferometer Gravitational-Wave Observatory (LIGO) to detect the feeble gravitational messengers generated by the most violent upheavals in the universe, including exploding stars and colliding black holes. Such hugely powerful astronomical events are required

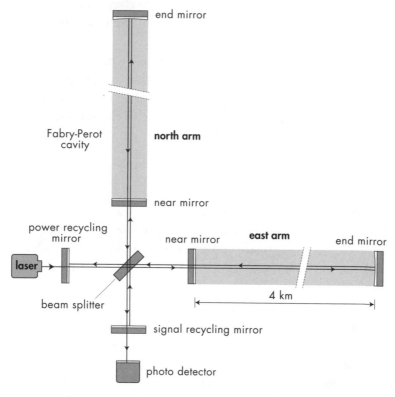

Schematic of the two arms of a LIGO detector to record gravitational waves.
(© Wil Tirion.)

to generate an appreciable gravitational wave because space-time, although elastic, doesn't easily flex or quake. Physicists calculate that space-time is a billion trillion times stiffer than steel.

LIGO's L-shaped arms serve as twin highways for a single laser beam, split in two. The tubes are long enough that Earth's curvature would bend each arm downward by a full meter; concrete and leveling keep the arms perfectly straight.

In the absence of a gravitational wave, the two laser beams travel exactly the same distance along each arm. The mirrors are configured in such a way that when the laser beams travel equal distances and then recombine, the crest of one light wave meets the trough of another and the two cancel each other out. The dark pattern of the recombined beams is known as a destructive interference pattern.

When a gravitational wave passes by, it alternately squeezes one arm while expanding the other. The two laser beams no longer travel exactly the same distance. This time when the two beams meet up, the crest of one light wave and the trough of the other are no longer in sync. Instead of canceling, the recombined light waves form a bright pattern whose duration and intensity reveal key details about the origin of the gravitational messenger.

The greater the distance the laser beams travel before recombining, the more sensitive the device is to the vibrations induced by a gravitational wave. Even four kilometers is not long enough to sense the tiny changes in length induced by a wave, so LIGO scientists employ a trick. Highly polished mirrors that absorb only one particle of light out of 3 million reflect each beam back and forth some 280 times before they merge, so that the distance they travel before recombining is 1,120 kilometers.

LIGO in fact requires two widely separated installations to detect gravitational waves—in addition to the Livingston site, another setup lies 3,000 kilometers away in Hanford, Washington. The two installations are necessary so that local, spurious vibrations at one

The LIGO installation at Hanford, Washington. *(Caltech/MIT/LIGO Laboratory.)*

site—a passing car, for example—are not mistaken for a gravitational signal from the heavens. Only a gravitational wave could create vibrations of equal strength within nanoseconds of each other at Hanford and Livingston.

Ten times more sensitive than an earlier LIGO version that operated at Hanford and Livingston from 2002 to 2007, Advanced LIGO detected its first gravitational wave on September 14, 2015, while the equipment was still being tested.

In August 2017, the Virgo interferometer, a gravitational wave detector jointly funded by France and Italy and based in Pisa, teamed up with LIGO and for the first time revealed the exact location on the sky of a gravitational wave.

In a cave dug deep within Mount Ikeno in Japan, engineers recently completed construction of the first underground interferometer to search for gravitational waves. Shielded from Earth's own vibrations and the noisy human world above ground by hundreds of meters of rock, the Kamioka Gravitational Wave Detector (KAGRA) joins other physics experiments in this quiescent subterranean lair. KAGRA's laser will travel along twin three-kilometer-long arms, reflecting off sapphire mirrors cooled to twenty degrees above absolute zero to further minimize spurious vibrations.

By the end of the next decade, a fifth gravitational wave detector, LIGO India, should be up and running. The first Asian outpost of LIGO, the interferometer will also feature twin four-kilometer-long interferometer arms like the ones at Hanford and Livingston.

However, ground-based interferometers have their limits. They can't detect gravitational waves with frequencies that fall below ten cycles per second because the random seismic vibrations of Earth are too large at those frequencies, hopelessly swamping the signal from space-time quakes.

To listen in to these longer wavelength ripples, the European Space Agency is now leading an effort to launch a detector called LISA (Laser Interferometer Space Antenna). Scheduled to fly in 2034, LISA consists of three spacecraft forming an equilateral triangle, with each side 2.5 million kilometers in length. Highly reflective, freely floating test masses within each satellite define the length of each side. Six laser beams traveling back and forth between the test masses would search for gravitational waves that would ever so slightly stretch or squeeze the spacing between the satellites. Scanning the entire sky as it trails Earth in its orbit about the Sun, LISA aims to measure relative shifts in position that are less than the diameter of a helium nucleus over a distance of 1.6 million kilometers.

A test mission, launched in December 2015, used the interior of a single spacecraft as a single, thirty-five-centimeter-long arm. The mission successfully demonstrated some of the advanced laser technologies needed for the full-blown mission.

The longer wavelengths LISA will be sensitive to are associated with larger objects and pairs of objects that are in much wider orbits and are far heavier than LIGO can record. With LISA, astronomers will be able to hunt for waves that come directly from the collision of supermassive black holes—millions to billions of times as massive as the Sun—that are believed to lurk at the center of nearly every galaxy. (In contrast, the stellar-mass black holes that LIGO detects are tens of solar masses.)

By recording the merger of supermassive black holes, astronomers can trace the merger of the galaxies in which these gravitational behemoths reside. Ultimately, LISA may reveal how little galaxies grew into larger ones throughout the history of the universe, setting the stage for the rich starlit tapestry spread across the cosmos.

Ultimately, LISA and more advanced detectors like it may be able to record the most fundamental of all gravitational waves—those that hark back to the birth of the universe. Astronomers believe that during the first tiny fraction of a second, the universe ballooned from the size of an atom to the size of a soccer ball. Following this mysterious and turbulent epoch, known as inflation, a cacophony of gravitational waves may have formed, and those waves may still exist today as a random background. The background would carry information about some of the universe's earliest moments, a few trillionths of a trillionth of a trillionth of a second after the Big Bang.

Astronomers have already detected relic light from the Big Bang. But that radiation, known as the cosmic microwave background (CMB), represents a snapshot of the universe when it was 380,000 years old. That's the era when the cosmos was cool enough for the universe to change from cloudy to transparent, allowing light for the first time to stream freely into space.

Before that time, light bounced back and forth in the hot dense soup of subatomic particles, and most—but not all—of the information that light carried about the Big Bang was erased. The leftover light could still bear the imprint of primordial gravitational waves, generated during inflation, which would have given the polarization of the CMB a particular curlicue twist.

At a March 2014 press conference at Harvard University that made headlines around the world, researchers using a radio telescope at the South Pole announced that they had found such a twist. Luminaries in the audience at the press briefing included many of the scientists whose theoretical work on the early universe would have been confirmed by the discovery. The discovery seemed worthy of a Nobel Prize. But the team later recanted, reporting that they had been fooled by cosmic dust particles, which can also twist the light in the same curlicue pattern. Teams continue to search for the imprint of these waves, which would confirm the theory of inflation.

....
....

DEEPER DIVE: Gravitational Waves Lost and Found

Two decades before Einstein proposed the existence of gravitational waves, Oliver Heaviside, a self-taught physicist and electrical engineer whose formal education ended at sixteen, beat him to the punch. Heaviside had no knowledge about ripples in space-time or curved geometry, but he was so enamored of a new theory about electricity and magnetism that he wanted gravity to fit in with the model.

According to the theory, which mathematical physicist James Clerk Maxwell brilliantly distilled from decades of observations performed by other scientists, electric and magnetic fields were intimately linked and were not stationary. Instead, they traveled freely through space as a single unit, called an electromagnetic wave, which moved at the speed of light. The wave acted to communicate the electric and magnetic forces over great distances.

Heaviside recast Maxell's equations into a more elegant format. And in 1893 he suggested that gravity should act in a similar way to electricity and magnetism—that it too should generate waves that traveled at the speed of light.

In 1905, the French philosopher and mathematician Henri Poincaré picked up the gauntlet, this time in the context of the newly emerging special theory of relativity, the mathematics for which he had begun developing independently of Einstein. The special theory holds that the speed of electromagnetic waves, which includes visible light, is a universal constant, measured to be the same value for all observers, no matter how slow or fast their relative motion. For this to hold true, no object or signal can ever travel faster than light, and observers moving at different relative speeds will see lengths contract and clocks slow down according to a precise mathematical recipe.

The theory explicitly embraces electromagnetic waves, which carry information about the forces of electricity and magnetism from one place to another. Poincaré reasoned that the same should hold true of all forces, including gravity. Thus, gravity should be communicated by gravitational waves—*ondes gravifiques*, he called them—that travel at the speed of light. Poincaré did not specify what the waves would look like or exactly how they would propagate.

Nonetheless, the idea of gravitational waves was revolutionary. It contradicted Newton's laws, as it implied there would be a time lag between any change in gravity and its effect on a distant object (dictated by the travel time required for a wave to be received by a distant body). According to Newton, gravity is communicated immediately—take away the Sun, and Earth would immediately fall out of orbit, as if some unseen being magically carried the change in force instantaneously across 150 million kilometers of space. (In the early 1800s, French scientist Pierre-Simon de Laplace suggested that gravity might propagate at a finite speed, but he did not consider the existence of actual gravitational waves.)

A decade later, when Albert Einstein had developed his general theory of relativity, portraying gravity as the curvature of space-time, he too suggested that there should be gravitational waves similar to electromagnetic waves. But for years he flip-flopped between believing in the existence of the waves and proclaiming his theory would allow for no such phenomenon.

In a letter he wrote on February 9, 1916, to German physicist Karl Schwarzschild, he declared, "There are no gravitational waves analogous to light waves." Only a few months later, using a simplified version of his complex equations governing gravity, he reported that he had found a wave-like solution similar to those in Maxwell's electromagnetic equations. But that calculation turned out to be in error. Still, in 1916, using a different set of coordinates (think of longitude and latitude on a map of the Earth), he found evidence for three kinds of gravitational waves. But as his colleague and eclipse observer Arthur Eddington noted in 1922, two of the waves could travel at any speed, even "the speed of thought," and were therefore spurious. The third wave, however, always traveled at the speed of light and seemed real.

In the 1930s, Einstein, who by then had fled Nazi Germany and set up shop at the Institute for Advanced Study in Princeton, again considered whether gravitational waves were real or not. In a draft paper written in collaboration with a young physicist, Nathan Rosen, he at first said the answer was no, but by the time the paper was published he had changed his opinion to a definite maybe. In any case, Einstein thought the waves would have such a tiny influence on any detector that they could never be measured.

A 1957 conference on general relativity, held in Chapel Hill, North Carolina, rekindled interest in gravitational waves, notably among one of the conference participants, Joseph Weber, a

University of Maryland engineer with steely determination and a wiry build. Weber designed and built two 3-ton aluminum cylinders, each designed to ring like a bell when a gravitational wave passed by. He placed one cylinder at the University of Maryland and the other at the Argonne National Laboratory near Chicago, 950 kilometers away.

To isolate the cylinders from spurious noise, Weber and his team suspended each one from a steel cable inside a vacuum chamber. A belt of detectors around the cylinders would record the vibrations induced by passing waves.

In 1969, Weber announced that his team had found gravitational waves, based on coincident vibrations of the two detectors. The report prompted research groups in Tokyo, Moscow, Munich, Glasgow, and the United States to build their own cylinder detectors. No one else ever found a signal, even though the new detectors were more sensitive. Moreover, calculations suggested that if Weber's findings were correct, our galaxy would be converting so much mass into gravitational waves that there would not be enough left over to keep the galaxy intact. To his dying day, Weber maintained that his results were correct, although everyone else had determined the signals were spurious. Regardless, his findings sparked a decades-long hunt for gravitational waves and helped make gravitational physics a mainstream endeavor.

In 1974, the search for gravitational waves got an unexpected boost. Russell A. Hulse and Joseph H. Taylor discovered a binary pulsar—two ultra-compact, rapidly spinning stars, orbiting each other at a furious rate. Each pulsar, which sweeps radio waves across the sky like lighthouse beams, crams as much mass as the Sun into a body the size of Manhattan. The orbital period of the pulsars—the time between the beacons of radio waves—is so steady that it varies by less than 5 percent over a million years.

That stability enabled Hulse and Taylor to look—indirectly—for evidence of gravitational waves. Einstein's theory of general relativity predicts that the orbital period of the pulsars, 0.05903 seconds, should ever so slowly decrease as the pulsars radiate away energy in the form of gravitational waves. The change is tiny, about 75 millionths of a second per year. But by the end of 1978, with four years of observations under their belt, Hulse and Taylor found an almost perfect match with Einstein's theory. The findings earned Hulse and Taylor the Nobel Prize in Physics in 1993 and were the first to give convincing evidence of gravitational waves.

It would require a new and more sensitive type of gravitational wave detector, using the laser interferometer, to directly detect gravitational waves in 2015.

8

IMAGING BLACK HOLES

AFTER A NIGHT of monitoring a worldwide array of radio telescopes from his office at Harvard, astronomer Shep Doeleman decided against the six-and-a-half-mile bike ride home and instead collapsed on the couch in the common area just down the hall. Around 5:00 A.M. his cellphone rang. Doeleman first thought the call might concern a technical glitch at one of the telescopes his team was operating, but no. Astronomers at the Large Millimeter Telescope—a huge radio dish perched atop a dormant volcano in a remote part of Mexico—had been run off the road by a group of men who pointed assault weapons at their heads.

Doeleman was used to dealing with problems and challenges at telescope sites—varying weather conditions, electrical outages. But, he thought, Astronomy 101 doesn't teach you what to do when members of your team have guns aimed at them.

Then again, nobody teaches you how to create a radio telescope with a diameter as large as Earth, just as nobody teaches you how to image the giant black hole at the center of the Milky Way. Those are skills you have to learn on the job.

Were the gunmen bandits, or were they undercover agents? Either way, they were *armed*. They had apologized and the astronomers had continued onto the mountaintop, but Doeleman and his colleagues decided that for the remainder of the ten-day observing run, the team would vacate the Large Millimeter Telescope and not return.

Then he got back to work. He still had seven other telescopes and arrays under his command, all staring at the same two targets—the supermassive black hole at the center of our galaxy and a heftier beast residing at the core of a galaxy some 50 million light-years away.

Black holes are invisible; any light that falls through their event horizon, the mysterious surface that surrounds every black hole and separates it from the rest of the universe, never gets out. Luckily for black hole hunters like Doeleman, the region just outside the event horizon blazes with light.

This radiation comes from two sources. One is the accretion disk, a swirling, doughnut-shaped shell of matter believed to orbit every black hole. As material from the disk spirals into the black hole at nearly the speed of light, it heats up to billions of degrees, prompting the matter to emit copious amounts of radiation. The other source is high-speed jets of glowing material expelled by black holes even as they draw material into their maw.

Doeleman and his collaborators have homed in on the radiation from the innermost part of the accretion disk, which lies nearest to the black hole. That's the light that most closely traces the shape and size of the scrambled region of space-time around the event horizon. Details of the image his team expects to see—a

Simulation of the region just outside the event horizon of a black hole, depicted at the radio wavelengths recorded by the Event Horizon Telescope. The bright ring comes from photons that orbit the black hole before reaching Earth and appears more luminous on one side due to the ring's rapid rotation. The shadow region inside the ring corresponds to photons that intersect the black hole and are swallowed by it instead of traveling to Earth. The shape and size of the ring provide a direct test of general relativity in the most extreme gravity environment in the cosmos. *(Courtesy J. Dexter, J. C. McKinney, E. Agol.)*

disk-shaped shadow limned by light—will put Einstein's theory to the ultimate test in the most extreme gravitational environment known.

The shadow is generated because the black hole swallows the light rays that come from directly behind it and which would otherwise have reached Earth. The warping of space-time by the black hole's massive gravity accounts for how light frames the shadow.

If the black hole were an ordinary object, the framing would be incomplete: the back portion of the luminous accretion disk, which lies behind the black hole, would be blocked from view. But the black hole's enormous gravity permits no such vanishing act. It distorts space-time so completely that light emitted from the back side of the disk, originally headed away from Earth, gets bent around so that it, too, reaches our planet. It's as if the black hole were immersed in a fog of light, says scientist Heino Falcke of Radboud University Nijmegen in the Netherlands.

This fog of light forms a halo that wraps around the entire shadow of the black hole. Unlike the prototypical angel's halo, however, the black hole's ring of light is not uniform in brightness. Because the gas in the inner part of the accretion disk orbits at nearly the speed of light, two different effects conspire to make the side of the disk rotating toward Earth appear brighter than the side rotating away. As a result, the halo appears more like a crescent. (See "Deeper Dive: A History of Illustrating Black Holes.")

To image the shadow, the team must record light at just the right wavelength. Much of the radiation from the innermost region of the accretion disk ricochets off the hot gas from the rest of the disk and never makes its way out. The light that does emerge must still travel through the fog of gas and dust cloaking the rest of the galaxy, which would blur the image at many wavelengths. At a radio wavelength of 1.3 millimeters, however, radiation from the inner part of the disk streams freely into space, easily penetrating gas and dust to complete the 26,000-light-year journey from the center of the Milky Way to Earth.

Calculations from scientists nearly a century ago had indicated that the two biggest shadows on the sky, from Sagittarius A* in the center of the Milky Way and the black hole at the center of M87, could be imaged from Earth. The shadows are big enough thanks to the gravitational distortion of each black hole on its own event horizon, which acts as a magnifying lens. The distortion enlarges the shadow by a factor of five, Falcke and his colleagues calculated.

Even then, the shadow would appear no bigger than 50 micro-arcseconds on the sky, about the same apparent size as a grape-fruit on the Moon. To discern such a shadow requires a telescope that can pick out features about two thousand times smaller than what the Hubble Space Telescope is capable of resolving.

The smallest structure that a telescope can see is defined by the wavelength of light divided by the diameter of the telescope. If the diameter of a radio dish is doubled, or the wavelength halved, a telescope can see objects twice as small. And that's when Doeleman and his colleagues realized that nature, for once, had been kind. For radio wavelengths longer than 1.3 millimeters, a tele-scope designed to image the shadow would have to be larger than Earth—a physical impossibility unless some of the instruments were aboard a spacecraft. At radio wavelengths much shorter than 1.3 millimeters, the telescope could be smaller and the image sharper, but most of the light would be absorbed by the Milky Way's gas and dust. At the Goldilocks wavelength of 1.3 millime-ters, the diameter of a radio telescope would have to equal that of Earth, but it would not have to be any bigger. And that size a black hole catcher Doeleman knew how to build. He wouldn't, of course, try to physically construct an Earth-size telescope. Instead, he'd use the properties of light waves to construct a virtual radio dish the size of the planet to peer into the hearts of black holes.

If a series of telescopes simultaneously detect light of the same wavelength from the same celestial source, the signals can be combined in such a way that the instruments act as a single telescope with a diameter equal to the largest separation, or base-line, between them. For instance, if two 4-meter-wide tele-scopes aimed at the same star are placed 100 meters apart, they can collectively discern the same amount of detail as a single, 100-meter-wide telescope.

The key is the rotation of Earth. If Earth were stationary, there would be no way that eight or more radio observatories spread across four continents could be linked to create a virtual telescope as big as the planet. Think of each observatory, Doeleman suggests, as a silvery speckle on a vast parabolic mirror, and the links between pairs of observatories as silvery lines. The silvered regions would not be enough to fill out the entire mirror. But as Earth spins, not only does each observatory see a different part of the sky, but the orientation between pairs of observatories changes. As a result, each silvery patch sweeps out a giant arc, creating the Earth-size instrument Doeleman's team needs to image black holes.

Telescopes working in concert rely on the interference between light waves—the pattern produced when two coherent light waves from the same source are superimposed—to create the resolution of a much larger instrument. A famous experiment first performed by English physicist Thomas Young at the turn of the nineteenth century illustrates the interference effect.

A beam of light is sent through two closely spaced slits, and the waves emanating from each slit meet on a distant screen. If the crest of one wave lines up with the crest from the other, they add together to make a stronger signal, forming a bright patch—a pattern called constructive interference—on the screen. If the two light waves combine so that the crest of one wave lines up with the trough of another, the two waves cancel each other out, forming a dark patch. From the intensity and spacing of the alternating pattern of bright and dark fringes on the screen, the shape and size of the object emitting the waves can be reconstructed.

In the observation Doeleman's team undertook, pairs of radio dishes take the place of Young's two slits. The radio emissions from the black hole's accretion disk propagate out through space as giant spherical waves, and when the wave front reaches Earth, each telescope detects a different part of that wave front. The radio waves will reach some telescopes earlier than others, since the instruments lie at slightly different distances from the galactic center. Even Earth's curvature and delays due to differences in the atmosphere at each site have to be taken into account. After the researchers compensate for the delays in arrival times, pairs of signals can be combined to form an interference pattern that can reveal the size and shape of the shadow.

Scientists had been using the radio telescope technique, known as very long baseline interferometry, or VLBI, for more than forty years. But no one had ever applied VLBI as Doeleman and his collaborators envisioned, to create a planet-wide telescope sensitive to radio emissions at 1.3 millimeters. And no one had ever routinely used VLBI to detect millimeter radio waves, which

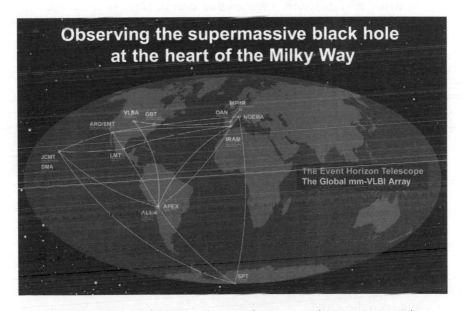

The worldwide network of radio telescopes that constitute the Event Horizon Telescope. *(ESO/O. Furtak/CC BY 4.0, https://creativecommons.org/licenses/by/4.0/legalcode.)*

required more precise timing and faster electronics than observations at longer radio wavelengths.

First, Doeleman had to convince the directors of several of
the world's largest radio observatories that their telescopes
should work in concert for this Event Horizon Telescope (EHT)
project, staring at the same celestial targets for several days
running each year. And once they agreed, he'd need to outfit each
site with atomic clocks and data recorders capable of storing millions of gigabytes of information a day—a rate higher than any
other science experiment ever attempted. In ten days, his team
would record more data than the world's biggest atom smasher,
the Large Hadron Collider in Switzerland, records in eight
months.

The project started with just a few telescopes. For two days in
April 2007, Doeleman and his colleagues commanded three observatories, two in North America and one in Hawaii, to record
radio waves from Sagittarius A*. Three observatories are not
nearly enough to image the shadow of a supermassive black hole.
But they sufficed to reveal that the telescopes had collectively
zeroed in on radio waves from a region just four times the size
of the black hole's event horizon—the diameter of Mercury's
orbit about the Sun.

No one had ever detected radiation from a region so close to a
supermassive black hole before. "That was the moment we really
thought we could do this," Doeleman recounted as he munched
carrots—part of a healthy, homemade lunch—in his Harvard office
the day after the armed encounter in Mexico. Doeleman, who has
a runner's build and looks a decade younger than his fifty-one
years, was still dressed in the biker shorts he wore to pedal to
work. "We were preparing for this for a while, we were thinking
about it, but seeing something this small, the size of the shadow,
set us on the path."

When interferometry is performed with visible-light telescopes, astronomers must combine in real time the signals recorded at each observatory to reconstruct the image. (There's currently no way for visible light waves to be stored for later comparison.) That means work on assembling a visible-light image can begin immediately and doesn't require a supercomputer. But that isn't possible with VLBI. The effort required to match up pairs of signals from the vast array of radio telescopes spread over the planet requires months of supercomputer time. And the amount of data recorded from each observatory is so great that the information can't be transmitted electronically.

To store the radio-wave signals, the EHT team in effect freezes the light from each observatory, creating faithful electronic copies of all the radio waves that fall in Chile, Mexico, Spain, Greenland, the South Pole, Hawaii, and Arizona. The time of arrival of every radio signal recorded at each of the telescopes and arrays must be recorded with high precision. That's the only way supercomputers can determine the difference in arrival time between radio signals recorded at each array, and identify which pairs of signals should be combined to create an interference pattern. Which is why Doeleman and his team installed at each observatory atomic clocks so precise they lose only a second every 10 million years.

"There are people who think deep thoughts with pencil and paper, and that is a big part of the scientific process," Doeleman told me in his office. "What excites me and has gotten me into this field is going to places and making new measurements and connecting wires together . . . to do something new that opens up a new window." Each year, new telescopes have joined the search. In 2017, the Event Horizon Telescope for the first time included a network of more than fifty radio dishes in Chile, the Atacama Large Millimeter/submillimeter Array (ALMA). Having ALMA

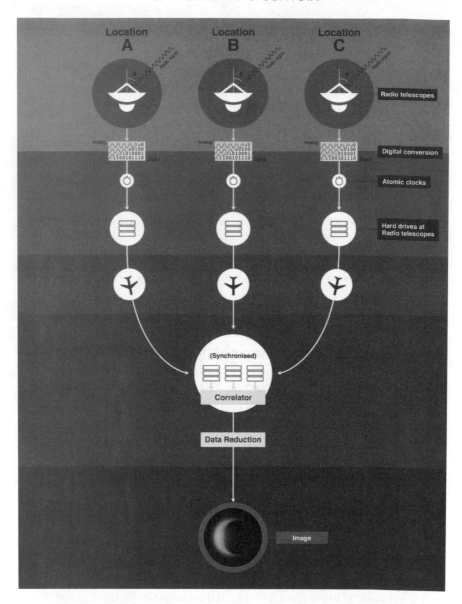

Schematic of the painstaking electronic method of combining observations from several radio telescopes to create a virtual telescope as large as Earth. *(ALMA (ESO/ NAOJ/NRAO), J. Pinto & N. Lira/CC BY 4.0, https://creativecommons.org/licenses/by/4.0/legalcode.)*

join the collaboration was a major coup; it improves tenfold the EHT's ability to see fine structure.

Even after installing and testing all the electronic equipment and meticulously planning out each observing run, the team must contend with something entirely unpredictable: the weather. Water vapor in Earth's atmosphere can absorb and emit the very same radio waves that the team depends on to image the shadows of Sagittarius A* and the black hole in M87, degrading the detected signals. Although the telescopes are placed at locations that are high and dry—mountaintops and desert plateaus—incoming snow, rain, and clouds laden with water vapor can threaten observations at any of the sites.

In April 2018, Doeleman's team had to choose the best five nights during a ten-day window to image the shadows, making the decisions on a day-by-day basis. (The team faced the same challenge during the previous run, in April 2017.) On a large whiteboard in Doeleman's office, his team kept a tally of the current and predicted weather at each site. Each observatory had its own row: the Submillimeter Telescope in Arizona, the Submillimeter Array and James Clerk Maxwell Telescope in Hawaii, the IRAM 30-meter telescope atop Pico Veleta in Spain, the South Pole Telescope, the APEX and ALMA arrays in Chile, the Greenland Telescope and the Large Millimeter Telescope in Mexico. The day after the astronomers were held at gunpoint, the team placed a red X next to the Mexico telescope.

The team typically made its go/no-go decision around 2:00 P.M. Eastern Time. Even before astronomers at the observatories checked in via phone calls, email, and videoconferencing, the team gathering in Doeleman's office scanned weather charts. Astronomers called in, talking of "tau"—radio telescope parlance for the opacity of the atmosphere. A tau of less than 1 means there's relatively little water vapor in the atmosphere to absorb the radio waves coming from Sagittarius A* and M87. A tau of 3 or higher spells trouble.

On that April day, the weather wasn't perfect—they had already decided that Pico Veleta would probably not observe that night—but the team knew that conditions would worsen at the two sites in Hawaii over the coming weekend, the last nights in that year's observing window. Doeleman bit the head off a chocolate Easter bunny, one of the snacks on the conference table. The decision was made: that night the EHT would observe.

The observations went well, but the next two days were troublesome. Although the telescopes in Chile had clear skies, both ALMA and APEX had technical problems. ALMA had a total power failure from which it took hours to recover, and APEX had an instrument failure that kept it off the air the entire time.

After each observing season, the data stored at each radio observatory—equivalent to the storage capacity of ten thousand laptops—are shipped to the Event Horizon Telescope's processing centers at the Massachusetts Institute of Technology's Haystack Observatory and the Max Planck Institute for Radio Astronomy in Bonn, Germany. Each year, Doeleman's team must wait patiently until the harsh winter ends in Antarctica and it's safe for a plane to fly out from the South Pole Telescope. At the two processing centers, stacks of supercomputers known as correlators search for and match pairs of radio signals that were emitted from the vicinity of the black hole at the same time but arrived at different times at each radio telescope. Combining such pairs yields "fringes"—the interference patterns required to image the shadow.

From the April 2017 data already in hand, Doeleman knew that his team had found fringes from celestial targets—bright quasars—that act as calibrators. That boded well for finding the fringes in the signals from the black hole, and indeed the 2017 data proved to be key to producing the first images of the event horizon of a supermassive black hole in 2019.

Exactly a century earlier, astronomers had struggled to record the bending of starlight from the Sun. Now they had recorded the bending of light within spitting distance of a black hole's event horizon—an observation that marked the beginning of a new century of discovery, one in which ghostly images will not only allow scientists to probe the nature of space-time around the strangest objects in the cosmos, but will continue to reveal precisely how well Einstein's theory of gravity describes the workings of the universe.

....
....

DEEPER DIVE: A History of Illustrating Black Holes

Growing up in a small town in the south of France, Jean-Pierre Luminet had plenty of time to draw, write, and paint. Using oil, pastel, and charcoal, he created portraits of famous composers like Liszt and Chopin and renowned scientists, including Newton. But when he was fifteen, Luminet's artistic leanings took a hairpin turn when he discovered the mathematically inspired optical illusions of the Dutch artist M. C. Escher and the nonsensical geometric figures of the English physicist Roger Penrose.

Drawings like Escher's staircase that seems to simultaneously descend and ascend and Penrose's perspective-shifting triangle seemed to defy Newton's laws of gravity. They captivated Luminet, who started sketching impossible architectures and wrong perspectives. Those drawings would prove to be a prelude for his life's work. While studying mathematics at Marseilles University in the 1970s, Luminet read an account of Einstein's general theory of relativity and began studying its mathematical underpinnings. After completing a doctoral thesis on a highly theoretical aspect of general relativity and joining Paris-Meudon Observatory in 1978, he followed the suggestion of his thesis advisor and tackled a more practical problem that would test his scientific and graphical skills: what would the accretion disk, the luminous, swirling platter of matter swaddling a black hole, look like?

Luminet was familiar with the archetypical images of black holes portrayed in the popular press: a dark sphere floating in a whirlpool of glowing gas. But for an accurate depiction, he would have to trace the paths that light rays are forced to follow around the gravitationally curved space-time surrounding a black hole.

The first efforts to seriously consider that question had actually begun a few years earlier, in 1972, when the physicist James Bardeen calculated the paths that light rays would take around a spinning black hole. Not long after, Bardeen and a former student, C. T. Cunningham, published the first image of what the position and intensity of a point source of light circling a black hole would look like to a distant observer over time. The image revealed that the black hole's gravitational bending of light lays bare the entire circular orbit, even the parts that lie behind the black hole.

To simulate in detail the appearance of a stationary black hole surrounded by a thin accretion disk, Luminet would have to perform a computer simulation. But before doing so, he used geometry to get a rough idea of what he would find.

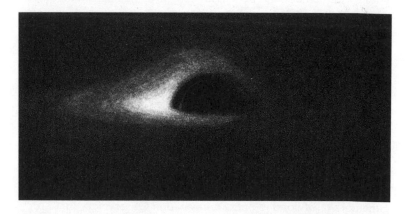

Jean-Paul Luminet's hand-drawn portrait of the region just outside the event horizon of a black hole, based on simulations using a primitive, transistor-based computer in 1979. The image compares remarkably well with more recent simulations using sophisticated computers. *(Courtesy J.-P. Luminet.)*

The situation, he realized, was very different from the appearance of another kind of accretion disk, the famous rings of ice and dust that surround Saturn. At any given moment, a portion of those rings is unseen as they pass behind Saturn. But around a black hole, the curved space-time forces light rays from the back side of the disk to bend around to the front, revealing the entire disk. More surprising is that light from the lower part of the disk, initially traveling downward and away from the observer, is redirected upward, so the upper and lower parts of the disk are also made visible.

Now Luminet was ready to write a computer program. At the Paris-Meudon Observatory, he used a primitive transistor-based computer, the IBM 7040 mainframe, which was built in the 1960s and required punch cards for inputs. Lacking drawing software, he created the final image by hand, using black India ink. In places where the simulation indicated light rays were concentrated, he peppered his imaging paper with a higher density of black dots.

His drawing resembled a halo that is brighter on one side than the other. The uneven luminosity is due to the rotation of the gas in the inner part of the accretion disk, which spins at nearly the speed of light. Just as the sound of an ambulance siren undergoes a change in pitch—a higher frequency as it approaches and a lower frequency as it recedes—light emitted by moving astronomical objects changes frequency. At any instant, the orbiting disk of light-emitting gas can be divided into two components: a part rotating toward an observer and a part rotating away. Light emitted by the approaching gas is shifted to bluer, shorter wavelengths while light moving away is shifted to redder, longer wavelengths. The fast rotation has a second effect: it concentrates the light in the direction of motion of the orbiting gas, so that the observed luminosity becomes brighter in some regions of the halo than others.

Over the years, other astronomers have performed more detailed simulations with much more powerful computers. Black hole physicist

Kip Thorne constructed the ultimate simulation for the 2014 movie *Interstellar*. Amazingly, that image is highly similar to Luminet's crude drawing twenty-six years earlier, though Thorne decided not to show the uneven brightness of the halo because he thought it would be too confusing for the audience.

SOURCE NOTES

Sources for direct quotes are listed below. For additional sources, please see Further Reading.

1. GENESIS

12 *"lazy dog"*: Quoted in G. J. Whitrow, ed., *Einstein: The Man and His Achievement* (London: BBC, 1967), 5.

12 *"Henceforth, space by itself"*: H. Minkowski, "Space and Time," in Hendrik A. Lorentz, Albert Einstein, Hermann Minkowski, and Hermann Weyl, *The Principle of Relativity: A Collection of Original Memoirs on the Special and General Theory of Relativity*, trans. W. Perrett and G. B. Jeffery (New York: Dover, 1952), 75.

13 *"happiest thought"*: Quoted in Abraham Pais,"*Subtle Is the Lord"*: *The Science and the Life of Albert Einstein* (New York: Oxford University Press, 1982), 175.

14 *"It is impossible to discover"*: Albert Einstein, *Out of My Later Years*, rev. repr. ed. (1956; New York: Citadel Press of Kensington Publishing Corp., 1984), 105.

DEEPER DIVE: SPACE AND TIME

21 *"superfluous learnedness"*: Quoted in Pais,"*Subtle Is the Lord,"* 152.

2. FROM TURMOIL TO TRIUMPH

30 *"As an older friend"*: Letter from E. G. Strauss to A. Pais, October 1979, quoted in Abraham Pais, *"Subtle Is the Lord". The Science and the Life of Albert Einstein* (New York: Oxford University Press, 1982), 239.

34 *"I know this way to the very end"*: Quoted in Jordan Ellenberg, *How Not to Be Wrong: The Power of Mathematical Thinking* (New York: Penguin, 2014), 395.

34 *"such wonderful things"*: Quoted in Ellenberg, *How Not to Be Wrong,* 395.
35 *"The assumption that the sum of the three angles"*: Stan Burris, "Gauss and Non-Euclidian Geometry," Crash Course Notes, September 2003, p. 12, translation of "Gauss to Taurinus, 8 November 1824," https://www.math.uwaterloo.ca/~snburris/htdocs/noneucl.pdf.
37 *"Grossmann, you've got to help me"*: L. Kollros, "Albert Einstein en Suisse: Souvenirs," *Helvetica Physica Acta* 29, no. 4 (1956): 278, quoted in Pais, *"Subtle Is the Lord,"* 212.
38 *"mere child's play"*: Albert Einstein and Arnold Sommerfeld, *Briefwechsel: 60 Briefe aus dem goldenen Zeitalter der modernen Physik von Albert Einstein und Arnold Sommerfeld,* ed. Armin Hermann (Basel: Stuttgart Schwabe, 1968), 26, quoted in John Earman and Clark Glymour, "Lost in the Tensors: Einstein's Struggles with Covariance Principles, 1912–1916," *Studies in History and Philosophy of Science, Part A,* 9, no. 4 (1978), 251.
40 *"I treat my wife as an employee I can not fire"*: Einstein to Elsa Löwenthal, Dec. 2, 1913, *The Collected Papers of Albert Einstein,* vol. 5: *The Swiss Years: Correspondence, 1902–1914* (English translation supplement), trans. Anna Beck (Princeton, NJ: Princeton University Press, 1995), doc. 488, pp. 364–365.

3. EDDINGTON ON A MISSION

48 *"The Germans are congenitally unfit"*: Sir Arthur Quiller Couch, quoted in Ariela Halkin, *The Enemy Reviewed: German Popular Literature through British Eyes between the Two World Wars* (Westport, CT: Praeger Publishing, 1995), 36.
52 *"I am a conscientious objector"*: Quoted in Matthew Stanley, *Practical Mystic: Religion, Science, and A. S. Eddington* (Chicago: University of Chicago Press, 2007), 146.
52 *"that question is not before us"*: Quoted in Stanley, *Practical Mystic,* 146.
52 *"My objection to war"*: Quoted in Stanley, *Practical Mystic,* 147.
53 *"a very hard one"*: Cambridge Daily News, June 28, 1918.
54 *"There is another point"*: Quoted in Stanley, *Practical Mystic,* 149.
59 *"Through cloud"*: The Observatory 540 (June 1919): 256.
60 *"Eclipse splendid"*: The Observatory 540 (June 1919): 256.
62 *"Today some happy news"*: Quoted in Abraham Pais, *"Subtle Is the Lord": The Science and the Life of Albert Einstein* (New York: Oxford University Press, 1982), 303.

63 *"The whole atmosphere"*: Alfred North Whitehead, *Science and the Modern World*, Lowell Lectures, 1925 (New York: Free Press, 1967), 10.

63 *"After a careful study"*: Quoted in Stanley, *Practical Mystic*, 110.

64 *"This effect may be taken"*: Quoted in Stanley, *Practical Mystic*, 110.

64 *"We owe it to that great man"*: Quoted in Pais, *"Subtle Is the Lord,"* 303.

64 *"This is the most important result"*: Quoted in Stanley, *Practical Mystic*, 111.

65 *"Revolution in Science"*: *Times* (London), November 7, 1919.

4. EXPANDING THE UNIVERSE

69 *"a burning question"*: Albert Einstein to Willem de Sitter, [before March 12, 1917], *The Collected Papers of Albert Einstein*, vol. 8: *The Berlin Years: Correspondence, 1914–1918* (English translation supplement), trans. Anna M. Hentschel (Princeton, NJ: Princeton University Press, 1998), doc. 311, p. 301.

70 *"The new version of the theory"*: Albert Einstein to Felix Klein, March 26, 1917, *Collected Papers of Albert Einstein*, vol. 8, doc. 319, p. 311.

71 *"The universe contracts into a point"*: Quoted in Eduard A. Tropp, Viktor Ya. Frenkel, and Artur D. Chernin, *Alexander A. Friedmann: The Man Who Made the Universe Expand*, trans. Alexander Dron and Michael Burov (Cambridge: Cambridge University Press, 1993), 157.

76 *"The evolution of the world"*: Quoted in Michael Rowan-Robinson, *The Nine Numbers of the Cosmos* (New York: Oxford University Press, 1999), 10.

77 *"Boys, we've been scooped!"*: Quoted in Alaina G. Levine, "Arno Penzias and Robert Wilson," *APS Physics*, 2009, https://www.aps.org/programs /outreach/history/historicsites/penziaswilson.cfm.

78 *"spherical bastards"*: Quoted in Kurt Winkler, "Fritz Zwicky and the Search for Dark Matter," *Swiss American Historical Society Review* 50, no. 2 (2014), 37.

78 *"dunkle Materie"*: Fritz Zwicky, "Die Rotverschiebung von extragalaktischen Nebeln," *Helvetica Physica Acta* 6 (1933): 110–127.

5. BLACK HOLES AND
TESTING GENERAL RELATIVITY

84 *"take this walk"*: Karl Schwarzschild to Albert Einstein, December 22, 1915, *The Collected Papers of Albert Einstein*, vol. 8, *The Berlin Years: Correspondence, 1914–1918* (English translation supplement), trans.

Anna M. Hentschel (Princeton, NJ: Princeton University Press, 1998), doc. 169, p. 164.

86 *"The star thus tends"*: J. R. Oppenheimer and H. Snyder, "On Continued Gravitational Contraction," *Physical Review* 56 (1939), 456.

7. HEARING BLACK HOLES

127 *"It sings! It laughs!"* "Edison's Phonograph," poster, ca. 1878, National Portrait Gallery, Smithsonian Institution, http://npg.si.edu/object/npg _NPG.87.225.

130 *"Stop what you're doing"*: David A. Coulter et al., Supplementary Materials for "Swope Supernova Survey 2017a (SSS17a), the Optical Counterpart to a Gravitational Wave Source," *Science* 358, no. 6370 (2017): S6.

133 *"We had no right to be so lucky"*: Author interview with Bernard Schutz, 2018.

DEEPER DIVE: GRAVITATIONAL WAVES

143 *"There are no gravitational waves"*: Albert Einstein, *The Collected Papers of Albert Einstein*, vol. 8: *The Berlin Years: Correspondence, 1914–1918* (English translation supplement), trans. Ann M. Hentschel (Princeton, NJ: Princeton University Press, 1998), doc. 194, p. 196.

143 *"the speed of thought"*: Arthur Stanley Eddington, "The Propagation of Gravitational Waves," *Proceedings of the Royal Society of London* A 102, no. 716 (1922), 269.

8. IMAGING BLACK HOLES

154 *"That was the moment"*: Author interview with Shep Doeleman, April 2018.

FURTHER READING

1. GENESIS

Chandrasekhar, Subrahmanyan. *Newton's Principia for the Common Reader.* Oxford: Clarendon Press, 1995.

Einstein, Albert. *Relativity: The Special and General Theory.* New York: Henry Holt, 1920. Reprint: Dover, 2010.

Einstein, Albert, and Leopold Infeld. *The Evolution of Physics: The Growth of Ideas from Early Concepts to Relativity and Quanta.* Cambridge: Cambridge University Press, 1961.

Galilei, Galileo. *Dialogue Concerning the Two Chief World Systems.* Translated by Stillman Drake. 2nd ed. Berkeley: University of California Press, 1967.

Holt, Jim. *When Einstein Walked with Gödel: Excursions to the Edge of Thought.* New York: Farrar, Straus and Giroux, 2018.

Holton, Gerald. *Thematic Origins of Scientific Thought: Kepler to Einstein.* Cambridge, MA: Harvard University Press, 1988.

Mahon, Basil. *The Man Who Changed Everything: The Life of James Clerk Maxwell.* Hoboken, NJ: Wiley, 2003.

Will, Clifford M. *Was Einstein Right? Putting General Relativity to the Test.* 2nd ed. New York: Basic Books, 1993.

2. FROM TURMOIL TO TRIUMPH

Antonick, Gary. "The Non-Euclidean Geometry of Whales." *New York Times,* October 8, 2012.

Bardi, Jason Socrates. *The Fifth Postulate: How Unraveling a Two-Thousand-Year-Old Mystery Unraveled the Universe.* Hoboken, NJ: Wiley, 2008.

Bonahon, Francis. *Low-dimensional Geometry: From Euclidean Surfaces to Hyperbolic Knots.* Providence, RI: American Mathematical Society, 2009.

Dunnington, G. Waldo. *Carl Friedrich Gauss: Titan of Science* (with additional material by Jeremy Gray and Fritz-Egbert Dohse). 1955. Reprint. Washington, DC: Mathematical Association of America, 2004.

Ferreira, Pedro G. *The Perfect Theory: A Century of Geniuses and the Battle over General Relativity*. Boston: Houghton Mifflin Harcourt, 2014.

Henderson, David W., and Daina Taimina. *Experiencing Geometry: Euclidean and Non-Euclidean with History*. 3rd ed. Upper Saddle River, NJ: Pearson, 2015.

Hoffmann, Dieter. *Einstein's Berlin: In the Footsteps of a Genius*. Baltimore, MD: Johns Hopkins University Press, 2013.

Holt, Jim. *When Einstein Walked with Gödel: Excursions to the Edge of Thought*. New York: Farrar, Straus and Giroux, 2018.

Isaacson, Walter. *Einstein: His Life and Universe*. New York: Simon and Schuster, 2007.

Levenson, Thomas. *Einstein in Berlin*. New York: Bantam Books, 2003.

Lorentz, H. A., et al. *The Principle of Relativity: A Collection of Memoirs on the Special and General Theory of Relativity*. New York: Dodd, 1923. Reprint: Dover, 1952. (Includes an English translation of Einstein's general relativity paper, which appeared in *Annalen der Physik* in 1916.)

Luminet, Jean-Pierre. *The Wraparound Universe*. Wellesley, MA: A. K. Peters, 2008.

Norton, John D. "Einstein's Pathway to General Relativity," Lecture for the class "Einstein for Everyone," University of Pittsburgh, https://www.pitt.edu/~jdnorton/teaching/HPS_0410/chapters/general_relativity_pathway/index.html.

Norton, John D. "General Relativity." https://www.pitt.edu/~jdnorton/teaching/HPS_0410/chapters/general_relativity/index.html.

Pais, Abraham. *"Subtle Is the Lord": The Science and the Life of Albert Einstein*. New York: Oxford University Press, 1982.

Weinstein, Galina. *General Relativity Conflict and Rivalries: Einstein's Polemics with Physicists*. Newcastle upon Tyne, UK: Cambridge Scholars Publishing, 2015.

Wheeler, John Archibald, and Kenneth William Ford. *Geons, Black Holes, and Quantum Foam: A Life in Physics*. New York: Norton, 2000.

3. EDDINGTON ON A MISSION

Crelinsten, Jeffrey. *Einstein's Jury: The Race to Test Relativity*. Princeton, NJ: Princeton University Press, 2006.

Douglas, A. Vibert. *The Life of Arthur Stanley Eddington*. London: Nelson, 1957.

Earman, John, and Clark Glymour. "Relativity and Eclipses: The British Eclipse Expeditions of 1919 and Their Predecessors." *Historical Studies in the Physical Sciences* 11 (1980): 49–85.

Kennefick, Daniel. "Testing Relativity from the 1919 Eclipse—A Question of Bias." *Physics Today* 62, no. 3 (2009): 37–43.

Stanley, Matthew. *Practical Mystic: Religion, Science, and A. S. Eddington.* Chicago: University of Chicago Press, 2007.

Will, Clifford M. "Henry Cavendish, Johann von Soldner, and the Deflection of Light." *American Journal of Physics* 56, no. 5 (1988): 413–415.

4. EXPANDING THE UNIVERSE

Belenkiy, Ari. "Alexander Friedmann and the Origins of Modern Cosmology." *Physics Today* 65, no. 10 (2012): 38–43.

de Swart, Jaco, Gianfranco Bertone, and Jeroen van Dongen. "How Dark Matter Came to Matter." *Nature Astronomy* 1 (2017): 0059.

Kragh, Helge, and Robert W. Smith. "Who Discovered the Expanding Universe?" *History of Science* 41 (2003): 141–162.

Mather, John C., and John Boslough. *The Very First Light: The True Inside Story of the Scientific Journey Back to the Dawn of the Universe.* Revised ed. New York: Basic Books, 2008.

Panek, Richard. *The 4 Percent Universe: Dark Matter, Dark Energy, and the Race to Discover the Rest of Reality.* New York: Houghton Mifflin Harcourt, 2011.

Tropp, Eduard A., Viktor Ya. Frenkel, and Artur D. Chernin. *Alexander A. Friedmann: The Man Who Made the Universe Expand.* Translated by Alexander Dron and Michael Burov. Cambridge: Cambridge University Press, 1993.

5. BLACK HOLES AND TESTING GENERAL RELATIVITY

Impey, Chris. *Einstein's Monsters: The Life and Times of Black Holes.* New York: Norton, 2018.

Thorne, Kip S. *Black Holes and Time Warps: Einstein's Outrageous Legacy.* New York: Norton, 1994.

Tyson, Neil deGrasse. *Death by Black Hole: And Other Cosmic Quandaries.* New York: Norton, 2006.

Will, Clifford M. "The Confrontation between General Relativity and Experiment." *Living Reviews in Relativity* 17 (Dec. 2014): 4.

6. QUANTUM GRAVITY

Carroll, Sean. "Does Spacetime Emerge from Quantum Information?" Blog post, May 5, 2015. http://www.preposterousuniverse.com/blog/2015/05/05 /does-spacetime-emerge-from-quantum-information/.

Lin, Thomas, Jennifer Ouelette, and K. C. Cole. "The Quantum Fabric of Space-Time." Three-part series in *Quanta Magazine*, April 2015. https://www .quantamagazine.org/.

Rovelli, Carlo. *Reality Is Not What It Seems: The Journey to Quantum Gravity*. New York: Riverhead Books, 2017.

Susskind, Leonard. *The Black Hole War: My Battle with Stephen Hawking to Make the World Safe for Quantum Mechanics*. New York: Little, Brown, 2008.

7. HEARING BLACK HOLES

Bartusiak, Marcia. *Einstein's Unfinished Symphony: The Story of a Gamble, Two Black Holes, and a New Age of Astronomy*. New Haven, CT: Yale University Press, 2017.

Collins, Harry. *Gravity's Kiss: The Detection of Gravitational Waves*. Cambridge, MA: MIT Press, 2017.

Levin, Janna. *Black Hole Blues: And Other Songs from Outer Space*. New York: Knopf, 2016.

Schilling, Govert. *Ripples in Spacetime: Einstein, Gravitational Waves, and the Future of Astronomy*. Cambridge, MA: Belknap Press of Harvard University Press, 2017.

DEEPER DIVE: LIGO AND BEYOND

Bartusiak, Marcia. *Einstein's Unfinished Symphony: The Story of a Gamble, Two Black Holes, and a New Age of Astronomy*. New Haven, CT: Yale University Press, 2017.

Keating, Brian. *Losing the Nobel Prize: A Story of Cosmology, Ambition, and the Perils of Science's Highest Honor*. New York: Norton, 2018.

DEEPER DIVE: GRAVITATIONAL WAVES LOST AND FOUND

Hunt, Bruce. 2012. "Oliver Heaviside: A First-Rate Oddity." *Physics Today* 65, no. 11 (2012): 48–54.

Kennefick, Daniel. "Einstein versus the *Physical Review*." *Physics Today* 58, no. 9 (2005): 43–48.

Kennefick, Daniel. *Traveling at the Speed of Thought: Einstein and the Quest for Gravitational Waves*. Princeton, NJ: Princeton University Press, 2007.

Rothman, Tony. "The Secret History of Gravitational Waves." *American Scientist* 106, no. 2 (2018): 96.

8. IMAGING BLACK HOLES

Luminet, Jean-Pierre, 2018. "Seeing Black Holes: From the Computer to the Telescope." *Universe* 4, no. 8 (2018): 86, https://doi.org/10.3390/universe 4080086.

DEEPER DIVE: A HISTORY OF ILLUSTRATING BLACK HOLES

Luminet, Jean-Pierre. "45 Years of Black Hole Imaging (1): Early Work 1972 1988." Blog post, March 7, 2018. https://blogs.futura-sciences.com/e-luminet /tag/black-hole/.

ACKNOWLEDGMENTS

It would not have been possible to write this book without the invaluable expertise, support, and patience of Richard Panek and Pamela Pagliochini. The author also expresses deep appreciation to the writer and astronomer Steve Maran, to Jeff Dean at Harvard University Press for his guidance and editing, to Louise Robbins, Emeralde Jensen-Roberts, and Stephanie Vyce at HUP, and to Sherry Gerstein of Westchester Publishing Services.

The author thanks the many scientists and science historians who provided help, insight, and advice, including Dieter Brill, Shep Doeleman, Heino Falcke, Vincent Fish, Daniel Harlow, Daniel Holz, Scott Hughes, Ted Jacobson, Daniel Kennefick, Jean-Pierre Luminet, John Mather, Charles Misner, John Norton, Don Page, Lenny Susskind, Brian Swingle, Virginia Trimble, Mark Van Raamsdonk, Galina Weinstein, and Clifford Will. Thanks also to Barbara Wood, Lorraine Wodiska, the Wednesday night league, Phil McQueen, and the eminent science writer and punster Davide Castelvecchi. And a special shout-out to Cathy Winter and Julie Cowen for their patience and support.

INDEX

Italicized page numbers indicate illustrations.